# Classical
# Mathematics

T0159019

*Joseph Ehrenfried Hofmann*

# Classical Mathematics

*A Concise History of Mathematics in the Seventeenth and Eighteenth Centuries*

Welcome Rain Publishers
New York

# Contents

CHAPTER ONE

# The High Baroque Period
# (approximately 1625-1665)

*Nec minor est virtus, quam quaerere, parta tueri.*
Ovid, Ars Amandi, lib. II, 13.

## 1. Descartes (1596–1650)

In contrast to the preceding periods of development which
were characterized by the coexistence of great numbers of indi-
vidually gifted personalities within the same type of environ-
ment, the mathematics of the High Baroque period was marked
by the emergence of several leading spirits of genius, who, in the
space of a few decades, brought about a complete change in the
interpretation of science. It now became a matter, not merely of
improving practical methods of application, but rather of estab-
lishing a comprehensive systematic structure of scientific
thought, to be constructed with mathematics as its model. To
make this possible it was necessary, first of all, to transform the
mosaic of conceptions arrayed starkly side by side into a self-
contained system determined by consistent tendencies.

Even for Copernicus (1473–1543) the point in question was
the construction of a new system of astronomy and, indeed, of
revitalizing and transforming ancient notions. Kepler (1571–
1630) abandoned tradition as a basis and followed intuition at
the crucial point. Plausible conjectures made on the grounds of

metaphysical speculation formed the starting point while the justification of these conjectures was tested and resolved most rigorously by comparison with observed data. In Galileo (1564–1642) the battle of contemporary physics against the basic tendencies of Aristotelianism came to the fore prominently. We shall meet with like viewpoints in R. Descartes (1596–1650) the highly gifted creator of the first modern system in the field of mathematics.

Descartes came of an old Norman family of nobility. His father, Joachim Descartes (d. 1640) was a councillor in the Breton parliament at Rennes. The boy, frail and motherless soon after he was born, was lovingly cared for (1604–12) in the newly established Jesuit college at La Fleche. There he received a painstaking and comprehensive education in science, the course in mathematics being prescribed by the textbooks by Clavius (Euclid edition of 1574, *Geometria Practica* 1604, *Algebra* 1608). As he was the second son of a well-to-do and highly respected family, it was intended that he should prepare himself for one of the higher public offices or for a military command. In 1613, he was presented at court, but finding that life in the world of Society was not suited to his tastes, he soon withdrew from it. In 1614, he began the study of law in Poitiers—although he had no deep inclination toward it—and there, in 1616, he was granted the degrees of *Baccalaureus* and *Licentiate*. In 1617, as a volunteer for active duty, he joined the Protestant army of the celebrated Maurice of Nassau, greatest general of his time. Descartes's friendship with the mathematician and natural philosopher, I. Beeckman (1588–1637) began in the encampment before Breda. Under the command of Duke Maximilian I of Bavaria and of Bucquoy, he took part in the expedition against the Winter King, Frederick V of the Palatinate. In Ulm, he came in contact with the master arithmetician, Faulhaber, and in the army winter quarters before Neuberg on the Danube, he discovered one of the principal ideas of his philosophy. It was in this period that his first independent discovery in mathematics was made, namely, the so-called Euler theorem on polyhedrons $e + f = k + 2$.

Weary of army camps, Descartes left the military service shortly after the Battle of the White Mountain (Nov. 8, 1620) and the occupation of Prague. He enlarged his knowledge of the world and of men through extended travel over Hungary, Germany and Italy (1621–25). Returning to Paris once again, he became an intimate friend of Mersenne (1588–1648) whose scientific friends were taking a stand against Aristotelian natural philosophy with visible success. Descartes found lively interest in this circle with respect to his novel points of view. He stayed for a short time at the camp near the Huguenot base, La Rochelle, where he became acquainted with the military engineer, Desargues. Then, in 1628, in a sudden decision, he returned to the Netherlands. He worked there in almost completely withdrawn solitude, on the formal construction of his system, keeping in touch only with Mersenne, out of the whole scientific community of Paris. In 1631, he went to England for a brief period; in 1634, to Denmark. He engaged in a correspondence on philosophy (from 1643) with Princess Elizabeth (a daughter of the Winter King) and it was for her sake that he went to Paris in 1644, 1646 and 1648. In 1644, he made the acquaintance of Chanut, the French ambassador in Stockholm. In 1647, the latter was the intermediary in the correspondence between Descartes and Queen Christina of Sweden. In 1649, Descartes accepted an invitation to visit the Queen, who wished to become familiar with his philosophy. Christina had under contemplation the establishment of an academy of sciences but Descartes died before the plans for the academy were carried out. His literary remains were brought to France by Chanut. Though the papers suffered some damage in a shipwreck, thanks to the efforts of Clerselier, most of those that were saved did reach publication (from 1657).

Under the influence of Ramée and Montaigne, Descartes broke away from traditional natural philosophy. It was, in his view, destined to remain barren, having become lost in inorganic enumerations, classifications empty of content, and interpretations wide of the mark. His goal was a method of investigation which, through an orderly arrangement of the subject matter, would effectively carry a well adjusted series of conclusions from the complicated to the simple and from conjecture to clarity and certainty. Mathematics, rigorously and systematically constructed, was the archetype for Descartes. Yet he was bent neither upon the search for the hidden methods of the ancients, nor upon the reconstruction of the substance of their writings which were known only by intimation. On the contrary, he wished solely for more precise definitions of concepts and the greatest possible generality in notation and methods of reasoning. His principal aim was the fusion of geometry and algebra, a goal which had been undertaken but not completely attained by Vieta. The technical terms to be employed were to be simple, clear and distinct, the method of notation was to be uniform, capable of being imprinted on the memory, and comprehensive. Thus, he replaced the mixture of words written in full, abbreviations and symbols appearing in the works of Vieta and his followers, by a pure symbolism so carefully thought through that it has endured almost unchanged to the present time.

Even before 1628, Descartes was acquainted with everything in mathematics that was concerned with order and measure. In the *Géométrie* of 1637, following the introduction of a unit line segment, he viewed every line segment as the representation of a number (arithmetization of geometry). Thus, by breaking through the principle of homogeneity hitherto regarded as inviolable on geometrical grounds, he was enabled to express the sum, difference, product and quotient of any two line segments as another line segment. As a result, he could also express line segments obtained by a combination of these four fundamental operations in a finite number of single steps. The number concept, originally restricted to natural numbers and extended to fractions, negative numbers and irrationals by the most laborious step after step, henceforth embraced the entire domain of algebraic numbers. Accordingly, Descartes considered problems solvable by purely algebraic methods (so-called geometrical) as belonging to exact mathematics, and all others (so-called mechanical) as belonging to the mathematics of approximation, which he no longer included in pure mathematics.

Descartes knew that all the geometrical problems of the linear and quadratic types could be constructed by straight edge and compasses and he classified such problems as plane problems (designation after the manner of Apollonius). In agreement with the cossists, problems of the 3rd and 4th degrees were called solid problems. Descartes solved the solid problems graphically (1628–29) by drawing a single parabola and cutting it with a circle which could be determined by plane constructions. Ferrari's method (see I, p. 84) for the solution of the 4th degree equation $x^4 + px^2 + qx + r = 0$ was transformed by means of $2t - p = y^2$. After the solution of the auxiliary equation $y^6 + 2py^4 + (p^2 - 4r)y^2 - q^2 = 0$, the value of $x$ was found from $x^2 \mp xy + y^2 + p/2 \pm q/2y = 0$. In accordance with his usual practice, Descartes gave only the prescribed formula for the calculation. His contemporaries verified it at the expenditure of a great deal of effort (Debeaune, printed 1649; van Schooten, printed 1659).

Descartes conceived of the graphical solution of equations of

higher degree by means of algebraic curves which he generated
by a step-by-step method using linear operating mechanism. He
viewed (in an unsuitable way) equations of the $(2n - 1)$th and
the $2n$th degrees as types of problems of the same $n$th class—
perhaps in the hope of the existence of the algorithmic solvability
of higher problems similar to the situation in the treatment of
solid problems. Descartes also stated theorems concerning the
general equation of the $n$th degree. By 1628, he had distinguished
between true (=positive) roots and false (=negative) roots. In
1637, he spoke of equations of the $n$th degree with $n$ real roots
and took into consideration the possibility of imaginary roots.
However, he did not let it be unmistakably known where he
stood on that most important discovery by Girard (1629: see I*,
p. 109), the fundamental theorem of algebra. He knew that
every integral root of an ordered equation beginning with $x^n$,
having only integral coefficients, is a divisor of the constant
term. He also used the divisibility of $f(x) - f(x_0)$ by $x - x_0$. The
heuristic method may have been used for the discovery of the
rule of signs, stating that the number of positive roots of an
equation is equal at most to the number of variations in sign
and the number of negative roots is equal at most to the number
of permanences in sign (more rigorous proof by Gauss, 1828).

These algebraic parts of the *Géométrie* were organically bound
up with far-reaching geometrical procedures.

To begin with, Descartes established that every algebraic
problem could be solved by the construction of certain line segments. Then he investigated the formulation of indeterminate
questions by connecting several unknowns in a system of equations fewer in number than the number of unknowns. The
locus *ad quattuor lineas* served as an illustrative example. In

---

* All references to the first volume in this series, "The History of Mathematics,"
by J. E. Hofmann, are given as I, p—.

1631, Gool had called Descartes's attention to this problem from Pappus. It was considered a difficult problem by contemporary mathematicians. The object was to find the locus of a point $P$ from which line segments $AP$, $BP$, $CP$, $DP$ may be drawn in any prescribed manner whatsoever, relative to four straight lines $a$, $b$, $c$, $d$, so that $AP \cdot BP = CP \cdot DP$. Descartes chose a fixed origin $O$ on the line $a$, set $OA = x$, the obliquely placed "applicate" $AP = y$. Then, by means of a relationship based on similarity, he expressed $BP$, $CP$, $DP$ as linear functions of $x$ and $y$. He eliminated the constant term of the resulting equation by a suitable choice of $O$ and obtained $y = p - mx + \sqrt{p^2 + qx + nx^2}$. He constructed and discussed the resulting curve, established the type of the conic section from the coefficients and also recognized the pair of straight lines occurring in the case of an exact square root. This was followed by an investigation of the complicated "locus to several lines" and also of the curves generated by mechanical geometric movement in which the reference line segments were taken in special positions of the greatest possible suitability. Occasional use was made of negative applicates; none, of negative abscissas.

Avoiding infinitesimal (hence, "approximate") analysis, Descartes solved the problems of normals to an algebraic curve as follows (purely algebraic method). He set the applicates perpendicular to the axis, took the point $P_0$ $(x_0, y_0)$, on the curve, the point $M$ $(t, 0)$ on the axis and let the circle whose center was $M$ and which passed through $P_0$ intersect the curve again at $P(x, y)$. Then he substituted the value $y = x - t + \sqrt{(x_0 - t)^2 + y_0^2}$ in the equation of the curve and obtained the equation $f(x, t) = 0$ with $x$ as the unknown, $t$ as the parameter, having $x - x_0$ as a linear factor. Now, Descartes required that the factor $x - x_0$ be split off again. Thus he found $t$ and from this, the normal $MP_0$. The conchoids $(x - c)^2 (x^2 + y^2) = a^2 x^2$ and the so-called Cartesian ovals were among the illustrative examples. A challenge was issued in the correspondence (from 1637) for the tangents to the *folium Cartesii* $x^3 + y^3 = axy$.

In an overestimation of his achievement, Descartes believed that by his method, he had made it possible to master all fundamental questions of exact mathematics. He regarded his method of normals and the method of tangents which flowed from it as the simplest methods possible. Undoubtedly he (together with Fermat) was the discoverer of coordinate geometry; whether he knew of the symbolical method, original with Nicole Oresme (see I, p. 63) is not certain. Yet, one would not be justified in designating Descartes as the discoverer of the analytical geometry of to-day. The concepts at the core of this science belong to a period not earlier than the beginning of the 19th century. Descartes was the most important individual among those who broke a path for the method of far greater scope. The greatest of his achievements was the grandiose algebraic structure of exact mathematics. However, he lacked insight into the deeper significance of transcendental questions although he did, indeed, have an excellent grasp of methods of solution for some of these problems in special cases.

As early as 1628, in connection with the discovery of the law of refraction, Descartes found the ovals having the equation $k_1 r_1 + k_2 r_2 = c$ in bipolar coordinates. These ovals, named after him, solved the problem of determining the surface of division between two homogeneous optical media, so that rays emitted from a point source of light in one medium may become re-united at a point in the second medium. During his investigation of free fall (1629) Descartes was misled (like Galileo, 1604) by a diagrammatic interpretation of the phenomenon of motion, and he arrived at the incorrect relationship $v = gs$ (instead of

$v = gt$). There was a mention of indivisibles here, but they were not accepted generally. In 1637, Descartes took up the Aristotelian dogma of the lack of comparability between the straight line and the curved line (see I, p. 20) stated in the form, no algebraic curve can be rectified algebraically. As evidence to the contrary, he offered the rectification of a transcendental curve, which could be carried out algebraically, namely, the logarithmical spiral. He defined this curve as the isogonal trajectory of rays through a point (1638). The period to which the posthumous "rounding out" of a given square belongs is uncertain. This was an isoperimetric transformation of the square into a regular polygon of 8, 16, . . . sides. Nor is it clear to what extent this work could have been instigated by similar undertakings by Nicolaus Cusanus (see I, p. 79).

Descartes also knew the construction for normals to the cycloid from the momentary pole and suggested how one could make the rolling of one curve upon another clear to himself through the use of systems of equal chords. Mention was made on this occasion, of the quadrature of the cycloid on the basis of the equality of sections of figures having equal altitudes (Cavalieri's principle: see I, p. 120). Shortly before this, Descartes gave, without proof, some quadratures, cubatures and determinations of centroids of plane parabolic surfaces and their solids of revolution "elucidated by the very statement of the results." He did not say how he had arrived at his results. Descartes was able (1638) to properly evaluate the quadrature of curves having the tangent property $\frac{dy}{dx} = \frac{x - y}{a}$ which had been achieved by Debeaune. He found the common asymptote $y = x - a$, but he failed in the attempt to find the curve itself (1639).

The *Géométrie,* the one mathematical work by the great philosopher, excited immediate attention, despite the unusual generality and intentional obscurity of its style. It was studied with interest. From 1638, copies of a commentary by an unknown author were in circulation among friends, and from 1639, the *Notae Breves* (printed 1649) in which the simpler single items of the *Géométrie* were demonstrated in scholastic form by the jurist, Fl. Debeaune (1601–52), was likewise circulated.

It was with the appearance of Fr. van Schooten (1615–1660) that Cartesian mathematics first penetrated into wider circles. As the son and successor of a highly respected university teacher at Leyden, van Schooten had received a first rate professional education.

From 1635 on, van Schooten was in personal contact with Descartes. During a journey to France, Ireland and England for the purpose of study, he became acquainted with the greatest authorities in mathematics at the universities of these countries and with their latest works, acquiring copies of mathematical manuscripts then in circulation (e.g. in Paris, copies of Fermat's treatises).

From 1645, van Schooten gave private instruction to numerous young Netherlanders who were interested in mathematical studies. He introduced them to selected classical works of ancient mathematicians and above all, to the methods of the *Géométrie*. In 1646, he published a voluminous collection of Vieta's writings which up to this time had scarcely become known, and, as the occasion offered, he replaced the unwieldy original style of notation by a transition to Cartesian symbolism. The public lectures of introduction to the *Géométrie*, given from 1646 on, were published in 1651, from a transcript of lecture notes worked out by Berthelsen. In 1649, the Latin translation of the *Géométrie* from van Schooten's pen, was issued after it had been examined by Descartes. It was followed by a detailed commentary and Debeaune's *Notae Breves*. In his *Exercitationes Mathematicae*, (1657; Netherlands, 1660) van Schooten also took up the most important individual results achieved by his pupils. The great two volume revised edition of the *Geometria* (1659, 1661) contained as supplements, Debeaune's studies on the theory of equations, Hudde's excellent works on extreme value methods and the determination of the limits of roots, de Witt's striking theory of conic sections based on instrumental methods of generation, beside numerous lesser corollaries and improvements in the explanations of the text, in which Huygens also took part.

This work which appeared in an edition having a large number of copies, was added to again, later on, with great success. It contained everything that from this time on was regarded as indispensable prerequisite knowledge for such persons as were interested in the newer mathematics and in the physics that was related to it. It superseded the earlier technical literature and it created a common foundation for the mathematical sciences of the Late Baroque period. Now, principle which to Descartes had been the only thing of importance, no longer received the greatest emphasis in the methods of thinking. The greatest prominence was given to operative points of view and to skillful handling of numerous individual examples. This work owed its great success, above all to this change in interpretation, whereas an abridged Netherlands treatment by G. Kinckhuysen (1660–63) met with less favor. Upon van Schooten's death, the Netherlands school collapsed. His insignificant successors remained immobilized in formalism. The interest of the next generation was devoted to the infinitesimal methods of richer prospect. The difficult basic problems of algebra (fundamental theorem, nature of the roots of an equation) were not to be undertaken again until the beginning of the 19th century, and then, on the basis of new points of view.

## 2. First Achievements in the Infinitesimal Domain
### (1629–1647)

We owe the first evidences of the more profound insights into
the domain of infinitesimals to the almost simultaneous studies
by Fermat, Roberval and Torricelli. Fermat, especially, pushed
forward to comprehensive general methods, but these were not
accepted at their true value by his contemporaries.

P. de Fermat (1601–1665), a native of southern France, came
of a respected middle class family. In his small home town,
Beaumont de Lomagne, he acquired wide literary and linguistic
knowledge. He studied law at Toulouse and served in the court
of justice, first as a lawyer, then, from 1631, as a member and
from 1634 as a councillor. We have learned of Fermat's mathematical achievements through his partially preserved correspondence, and of some of his methods through rather small
treatises, very few of which were printed during the lifetime of
the author. Most of his treatises were not printed until 1679
when they appeared in inadequate form. However, their contents
had become known earlier, for the greatest part, through copies
which were in circulation. An extremely busy man, Fermat was
given to making notes of ideas that came to his mind on slips of
paper which were later heedlessly laid aside, or to making nota-
tions in the form of marginal notes in books which he was using.
The celebrated marginal notes in his manuscript copy of the
Bachet edition of Diophantus (1621) were added to the second
edition of 1670.

Fermat desired to combine mathematical methods taken from
the ancients with those of contemporary mathematicians and to
realize the greatest possible rigor and generality. Before 1629, he
had found $\int_0^1 t^p \, dt$ ($p$ integral and positive) to be the quadrature

of a surface by passing to the limit with $n \to \infty$ using the recurrence formulas $2\sum\limits_{k=1}^{n}k = n(n+1)$, $3\sum\limits_{k=1}^{n}k(k+1) = n(n+1)(n+2)$, etc. (by letter 1636). Dissatisfied with this method, he discovered (about 1629) that the area $\int\limits_{0}^{a} y\, dx$ of the parabola $\frac{y}{b} = \left(\frac{x}{a}\right)^{p}$ could be found in general form by a fusion of the Euclidean determination of a pyramid (see I, p. 24) with the geometric Archimedean quadrature of the parabola (see I, p. 25) in which inscribed and circumscribed figures are constructed in steps so that the abscissas of their points of division form a geometrical progression on the X-axis (first term $a$, second term $t < a$; in the computation of the area of the entire infinite series of these step-figures, $t \to a$). A short time later he extended the method to the parabolas $\left(\frac{y}{b}\right)^{q} = \left(\frac{x}{a}\right)^{p}$ ($p$, $q$ integral, prime to each other and positive), to the spirals $\left(\frac{r}{a}\right)^{q} = \left(\frac{r\varphi}{a}\right)^{p}$ and $\left(1 - \frac{r}{a}\right)^{q} = \left(\frac{r\varphi}{a}\right)^{p}$ (mentioned in Mersenne's writings 1637–1644) and to the hyperbolas $\left(\frac{x}{a}\right)^{p}\left(\frac{y}{b}\right)^{q} = 1$. He knew that equal areas lay in this "logarithmic" division of the base line in the case of the hyperbola $xy = ab$ before passing to the limit. From 1636, there was added to this, the cubature and determination of the centroid for solids of revolution with parabolic meridian sections and the calculation of hyperbolic areas extending to infinity by the use of improper integrals.

The impetus for the extreme value rule (1629) came from Pappus VII (see I, p. 35): if in $\frac{f(x+h) - f(x)}{h}$ the coefficient of the lowest power of $h$ is set equal to zero, then a value of $x$ will be found for which $f(x)$ will assume an extreme value (necessary condition). The rule was intended at first only for polynomials. This is apparent as much from the original illustrative example (greatest rectangular area having a fixed perimeter) as from the later examples. However, in its well thought through formulation, it was valid generally, a fact which Fermat pointed out with great emphasis.

In later settings (1643) the discussion concerned passing to the limit with $h \to 0$ and the distinction of types of extremes arising from the sign of the term in the case of $h^2$. Irrationals were handled by means of rationalizing substitutions. Fermat based his method of tangents on the following procedure: He determined the subtangent $t$ of the curve $y = f(x)$ from $\dfrac{t - h}{t} \approx \dfrac{f(x - h)}{f(x)}$ (earliest example: the parabola $y^2 = 2px$). Using an original method, he found the abscissa $\xi$ of the center of gravity of the surface $\int_0^a y \, dx$ belonging to the parabola $\left(\dfrac{y}{b}\right)^q = \left(\dfrac{x}{a}\right)^p$ and of the corresponding solid of revolution $\int_0^a y^2 \pi \, dx$. In the case of the solid of revolution (earliest example: $y^2 = 2px$), Fermat set $\triangle \xi \cdot \int_0^{x - \triangle x} y^2 \pi \, dx \approx (x - \xi) \cdot \int_{x - \triangle x}^{x} y^2 \pi \, dx$ and applied the condition for homogeneity $\dfrac{\triangle \xi}{\xi} = \dfrac{dx}{x}$. He was, of course, also acquainted with the usual formula for the determination of $\xi$.

According to allusions made in his letters (1636–43) an early (lost) manuscript by Fermat on number theory was concerned with the sum $\Sigma a$ of the divisors of a number $a$ (including 1 and $a$). It contained rules for the treatment of composite numbers and iteration methods whereby certain quantities which had been constructed could then be shown to be prime numbers. With their help, Fermat solved $\Sigma a = 2a$ (perfect numbers) as Euclid had done (see I, p. 23) and in addition (like Descartes, 1636) he solved special cases of the general equation $\Sigma a = \lambda a$ ($\lambda > 1$, rational). The system $\Sigma a = \Sigma b = a + b$ (friendly numbers) was handled by Fermat in the manner of Tâbit (see I, p. 41) by means of $a = 2p_1p_2q$, $b = 2p_3q$, choosing $q = 2^n$ so that $p_1 = 3q - 1$, $p_2 = 6q - 1$, $p_3 = 18q^2 - 1$, would become prime numbers.

He observed that $2^p - 1$ would be prime at best if $p$ were prime and $2^m + 1$ likewise, if $m = q$. After the factorization of $2^{37} - 1$ (223 factors), he stated in general form that $a^{p-1} - 1$ was divisible by $p$ (so-called lesser Fermat theorem) and that in any case such prime factors of $a^p - 1$ as did exist took the form of $2kp + 1$. He conjectured erroneously, on the basis of incomplete induction, that $2^q + 1$ is always prime (counterevidence by Euler 1732: 641 is a divisor of $2^{32} + 1$).

In a corollary by Campanus to Euclid IX, 16, in which the irrational produced by continued division, $x^2 + xy = y^2$ was derived arithmetically from $x : y = (y - x) : x$, Fermat perceived the model example of a comprehensive method, his *descente infinie*. It demonstrated that the solvability of a problem could be carried out more effectively by falling back to ever smaller solutions, through the narrowing down of the intervals, than by the reverse—the method of complete induction—and that this was at the basis of the earliest infinitesimal studies of the ancients. After many futile attempts, Fermat, using this procedure, confirmed Girard's assertion (in print, 1634; see I, p. 102, 103) that every prime number in the form $4n + 1$ is the sum of exactly two squares making use of the reconstructed Diophantine identity $(a^2 + b^2)(c^2 + d^2) = (ac \pm bd)^2 + (ad \mp bc)^2$. Similarly he constructed the prime numbers $8n \pm 1$ by means of $2x^2 - y^2$, the prime numbers $2(4n + 1) \pm 1$ by means of $2x^2 \pm y^2$, and the prime numbers $6n + 1$ by means of $3x^2 + y^2$. Beside this, he decomposed $8n + 3$ into the sum of three squares, $8n + 7$ into the sum of four squares, he was aware of the related rules for decomposition and divisibility, and he represented every integer as the sum of $p$ $p$-gonal numbers (1638). These theorems were of service to him in the factorization of a given number based on $x^2 - y^2$, where he applied step-by-step addition.

Following in the footsteps of Diophantus, Fermat engaged in a detailed study of indeterminate problems concerned with integral right triangles having $x$, $y$ and $z$ sides, so that $x^2 + y^2 = z^2$. He established that the area $\triangle = \dfrac{xy}{2}$ could be neither $\square$ nor $2\square$, enunciated the rules for the determination of triangles having given ratios, constructed triangles of equal area and any number of triangles in which the sum or the difference of the sides including the right angle was given. By iteration processes which were skillfully adjusted to the individual problem, he mastered the difficult cases, (1643) somewhat like $(x + y)^2 + \triangle = \square$ ($z$ 7 digits) or $x + y = \square$, $z = \square$, ($z$, 13 digits). By his method of *descente*, Fermat discovered that there could be no solution to the problems in whole numbers, none of which is zero, which were referred to again and again in correspondence from 1640: (1) $x^4 + y^4 = z^2$, (2) $x^3 + y^3 = z^3$, (3) $p^2 - q^2 = q^2 - r^2 = r^2 - s^2$. His assertion, that beyond this, $x^p + y^p = z^p$ ($p > 2$, integral) is impossible (the so-called greater Fermat theorem), appeared only in marginal notes in an edition of Diophantus.

Through problems of this type, Fermat did indeed demonstrate his preeminence over his rivals who were working on tables and heuristic procedures; but he never achieved the forward-looking exchange of ideas that he hoped for with colleagues who were in some measure of equal rank. The most significant of his correspondents on number theory was the arithmetician B. Frenicle (1605–1675). Their correspondence began in 1640 with a controversy over magic squares in the course of which Fermat gave rules for the construction of squares such that they would remain magic after the removal of single-celled borders.

By 1629, Fermat had reconstructed the second book of the *Plane Loci* by Apollonius (see I, p. 29) on the basis of the wording of the propositions in Pappus VII. This was followed in 1636 by the much more interesting first book whose first *prop.* (inversion geometry of the circle) was resolved by Fermat into eight component theorems. The construction of the parabola passing through four points came a little earlier. At the same time, he proved by pure geometry that the locus to three lines $AP \cdot BP = CP^2$ was a conic section (cf. p. 8).

Toward the end of 1636, in an endeavor to uncover the full meaning of the ancient works on geometric loci, Fermat, like Descartes, of whose works he was completely unaware, hit upon the idea of the determination of a point in a plane by coordinate geometry. He did this with frequent use of perpendicular applicates. He gave the straight lines through the origin in the form $bx = ay$, and considered $ax + by = c$ as the equation of a straight line lying in a general position. He extended an Apollonian locus problem through the recognition of $\Sigma(a_ix + b_iy + c_i) = d$ as a straight line. Beside this, he constructed one of the straight lines determined by $ax^2 + bxy + cy^2 = 0$ (in the case, $b^2 > 4ac$). In addition, he interpreted $xy = a^2$ as an equilateral hyperbola, found the center and the asymptotes for $xy + c^2 = ax + by$ and gave the equations of the parabola referred to the vertex, and the equations of the circle, ellipse and hyperbola referred to the center. As an appendix, he reduced solid problems in $x$, by an appropriate choice of $y$, to the intersections of conic sections, and ultimately, to the use of the parabola and the circle. In this short treatise, the essential nature of the analytic method was displayed more clearly than in Descartes' work. In particular, clarity prevailed in the meaning of the sign preceding a term where coordinates were used. The presentation, it must be admitted, was cumbersome, as Fermat held to the ungainly style of writing due to Vieta. It was about the same time that the first theorems of a sketch on surface loci were originated. The sketch was completed in 1643. Fermat made an effort to identify surfaces of the second order by their plane sections, but he overlooked ruled surfaces.

Immediately upon the receipt of Descartes' *Géométrie* (1637), Fermat sent Descartes, through Mersenne as intermediary, copies of the extreme value rule and the introduction to the geometrical loci. This was done in order to establish the independence of his own methods securely. Descartes declared the tangent rule to be incorrect on the basis of a prejudiced and all too superficial examination. However, after he had received more detailed explanations from Fermat with excellently selected examples and supplementary remarks, he conceded his error. Unfortunately, he expressed himself scornfully concerning Fermat even later. For example, in a conversation with van Schooten in 1644, Descartes spoke of Fermat as a boastful Gascon. This opinion was taken up by the next generation of scientists whose skill in proof was inadequate as yet to render Fermat's results in number theory accessible to them. Thus, they erroneously regarded Fermat's achievements as the fruits of heuristic considerations, and consequently as accidental products without far-reaching significance. Because of this, Pascal (1654) and Huygens (1658) denied themselves to the aging Fermat and through the lack of time for the compilation of his methods in number theory, they prevented him from becoming a collaborator. Whatever was compiled out of Fermat's letters by the second rate J. de Billy (1602–1679) for an introduction to an edition of Diophantus was distorted by misprints which destroyed the meaning, and the excerpts were concerned with inconsequential special cases instead of general interrelationships. Among the correspondents, it was Mersenne who, though he was hardly a technical expert, endeavored to make Fermat's work more widely known. At first only handwritten copies of Fermat's treatises were in circulation. Mersenne took up individual results achieved by Fermat in his own publications 1637–38, 1644), with appropriate reference to the author. He was the intermediary in the reprinting of the tangent rule in the *Supplement* to Herigone's *Cursus* (1642) and in Fermat's relations with the Italians (from 1643). After Mersenne's death, Fermat availed himself of the services of his old friend, P. de Carcavy (1600?–1684), above all, as intermediary in communicating with foreign correspondents.

In the important correspondence (from 1636) between Fermat and the ambitious and arbitrary G. P. de Roberval (1602–1675), questions relating mainly to the infinitesimal were discussed. Roberval taught at *Collège Royal* from 1630, and from 1634 he also occupied the chair of mathematics established by Ramée. During the latter period, Roberval carried out numerous quadratures (independent of Cavalieri) through the application of indivisibles. In the quadrature of $\int_0^1 t^p \, dt$, he showed a preference for $p = 1$ and 2, as Stevin did (cf. I, p. 103). This result was extended by him without adequate foundation to all integral and fractional values of $p$. He determined, among other things, the area of the conchoid of a circle, the surface area of the cylinder $x^2 + y^2 = ax$ up to its intersection with the sphere $x^2 + y^2 + z^2 = a^2$ and the lateral area of an oblique circular cylinder.

As early as 1629, Mersenne directed Roberval's attention to

the cycloid $z = y + s$ (generated by the circle $x^2 + y^2 = 2ax$). In 1634, Roberval achieved the quadrature $\int\limits_{0}^{2a} z \, dx = \frac{3a^2\pi}{2}$ with the help of the area $\int\limits_{0}^{2a} s \, dx = a^2\pi$ of the "companion" to the cycloid $\overline{y} = s$. Fermat discovered a method of proof involving symmetry in 1638. By 1637, at the latest, Roberval was in possession of a tangent rule to which he was led by a study of the Archimedes treatise on spirals (cf. I, p. 26). He presented numerous well thought through examples among which was the determination of tangents to a cycloid. In 1638, Descartes added the construction of the normal from the momentary pole in the extension of the problem to prolate and curtate cycloids, and Fermat gave the geometrical construction of the tangents from $\frac{dz}{dx} = \frac{y}{x}$. Roberval found the point of inflection for the conchoid of a line in 1638, a noteworthy discovery. It led Fermat to the formulation of the general condition for a point of inflection. In 1643, Roberval related the Archimedean spiral $r = a\varphi$ and the parabola $x^2 = 2ay$ through the equality of arcs.

The most valuable discovery made by Roberval fell in the period around 1645 (by letter, 1646), namely, the intuitive interpretation, based on $\frac{dy}{dx} = \frac{y}{t}$ of the change of variable in integration, with $\int\limits_{a}^{a} y \, dx = \int\limits_{\beta}^{b} t \, dy$, by means of which he solved the quadrature of the parabola and hyperbola in general form, passing from the curve $(x, y)$ to the curve $(x - t, y)$.

Roberval's mathematical writings, conveyed into print for the first time out of his literary remains (1693), show that this richly creative investigator lacked power in formal organization. The inferiority complex evoked by this disparity between the will and the capacity to do, explains Roberval's animosity toward Descartes. In 1646, at a public meeting on the algebraic content of the *Géométrie*, Roberval accused Descartes of plagiarizing from Harriot and Oughtred (accusation seized upon by Wallis, 1685). This complex also explains his animosity toward other actual rivals or such as he presumed to be rivals, toward Torricelli, above all.

The Tuscan, E. Torricelli (1608–1647), aroused great hopes even as a student at the Jesuit college at Faenza. From 1627 on, under the tutelage of the Camaldolite Abbot Castelli, an enthusiastic follower of Galileo, he became familiar with the principal works in ancient mathematics and mathematics of more recent times. He was also introduced to the basic ideas of the modern tendency in natural philosophy. The rather small geometrical studies which he left behind, though written in his youth, demonstrate just as much as his treatises on the sphere and the quadrature of the parabola (printed 1644), that Torricelli had thoroughly mastered the Archimedean method and its development by Valerio (cf. I, p. 117, 118). He combined it successfully with Cavalieri's method of indivisibles (1635), and while he was well aware of the snares in the method, he also knew how to avoid them skillfully.

The very first independent work by Torricelli contained important new results. Referring to Cavalieri (1632; see I, p. 121), Torricelli investigated by pure geometry, the parabolic path of a projectile having a fixed initial velocity (1641, printed 1644). Here he presented the position of the vertex, the envelope and the point of tangency of the envelope.

During a short stay in Arcetri (1641–42), under the direction of Galileo who had now become blind, Torricelli undertook the final editing of the appendices (ed. Viviani 1674) to the *Discorsi* (determination of the center of gravity, proportionality). At this time, he did some original work on the same subject in which the formula for the centroid $\xi$: $(a - \xi) = \int_0^a xy \, dx : \int_0^a (a - x) \, y \, dx$ and (referring to Valerio) examples in spherical segments for the so-called Simpson rule (1743) also appeared (1642).

In 1641, Torricelli achieved the cubature $\int_a^\infty \pi y^2 \, dx = \int_0^b 2\pi y \, (x - a) \, dy = a \, b^2 \pi$ of the solid of revolution of the equilateral hyperbola $xy = ab$ by integration by parts (printed 1644). Following a suggestion from Viviani, he applied the mechanical tangent rule (found independently of Roberval) to the ordinary, the prolate and the curtate cycloids (1643–44), and he treated simple inverse tangent problems. In addition, he discovered the construction of the normal by means of the momentary pole, rectified the logarithmic spiral generated by $r = a^{-t}$, $\varphi = \alpha t$ and squared the ordinary cycloid which he conceived as the image of a helix on a circular cylinder, whose generating element would be cut at an angle less than $45°$. These results were independent of Descartes.

Mersenne was in contact with Galileo as early as 1635. His friend and fellow friar, Nicéron, was the intermediary in the letters between Cavalieri and Fermat concerning the quadrature of the parabola. In 1643, Roberval and Fermat received an account of the new results which Torricelli had just finished for publication. In 1644, Mersenne made a pilgrimage to Rome and while there, he gave Fermat's extreme value problems to Cavalieri and Torricelli. Both of these mathematicians solved the problems by direct geometrical reasoning (prettiest example: the point such that the sum of its distances from three given points is a minimum). Later, reports of other discoveries by the French mathematicians followed and copies of Fermat treatises, in particular, were translated into Italian by Torricelli's highly gifted pupil, the great linguist, Ricci.

In 1646, Torricelli combined Fermat's method for the quadrature of the parabola which was similar to his own of 1644 (his own was used, however, only for integral exponents), with the rectification of the logarithmic spiral (by chords and tangents in geometric series). He also indicated the tangent properties of the parabolas and the hyperbolas $\left(\dfrac{y}{b}\right)^q = \left(\dfrac{x}{a}\right)^p$ by $\dfrac{q\,dy}{y} = \dfrac{p\,dx}{x}$.

He achieved the quadrature by $\int_y^b x\,dy : \int_x^a y\,dx : (ab - xy) = p : q : (p + q)$ and subsequently, he stated the condition for the existence and the value of the improper integral $\int_a^\infty y\,dx$ for the case of the hyperbola. He set to work, correspondingly (1647) on the higher spirals $\left(\dfrac{r}{a}\right)^p = \left(\dfrac{r\varphi}{a\alpha}\right)^q$, whose polar subtangent $t$ he determined from $t : r = r\,d\varphi : dr = (p - q)r\varphi : qr$. By means of the transformation $r = x$, $r\varphi = y$ or, $r\varphi = py : (p - q)$, he reduced the quadrature and the rectification of the spirals to the corresponding procedures for the parabolas $\left(\dfrac{y}{b}\right)^q = \left(\dfrac{x}{a}\right)^p$. In the transition from the distance $s(t)$ to the velocity $v(t)$ and the reverse, he showed that the inverse character of the tangent problems and the quadratures had become clear to him. He knew, in addition, that if $x : p = y : q$, then $x^p y^q$ would be a maximum, assuming that $x + y = a$.

Beside these investigations, whose rigorous proofs in pure geometric form were well-nigh cut off by Torricelli's sudden death, there appeared a sketch of the logarithmic curve $x = a \log \dfrac{y}{b}$ having $a$ as the constant subtangent. Using $a\,dy = y\,dx$ Torricelli found the area extending to infinity between the curve and its asymptotes and the cubature of the corresponding solid of revolution (1646–47). Here, as in the treatment of simpler questions in geometry, he proved himself to be an ingenious investigator who would forego an algebraic expedient in favor of a pure method, and who, having a grasp of the fundamental problems of the infinitesimal mathematics of the time, handled them in a masterly fashion. Even though his most elegant results remained in manuscript form for hundreds of years, they were, nevertheless, influential, first in a narrow circle of friends and then, through Ricci and Angeli upon Sluse, Gregory, Barrow and Leibniz. Thus, they contributed decisively to the evolution of modern mathematics.

Among Torricelli's pupils and friends, Viviani and Ricci must be mentioned, above all. V. Viviani (1622–1703), a Florentine, was Galileo's last direct pupil. From 1639 on, he served as secretary to Galileo, who had already lost his sight. Fundamental studies of the ancients were his life work in mathematics. As early as 1642, he published an important restoration of Book V of the Apollonian *Conic Sections,* which, at that time, was still unknown in the original (printed 1659). The work was done on the basis of ten lemmas cited in Pappus VII and those alluded to in the Eutocius commentary on Archimedes. In 1645 he tried his hand at the restoration of the lost solid loci of Aristaeus (printed 1673).

The clergyman, M. Ricci (1629–1682) who was raised to the post of cardinal in 1681, came of an impoverished noble family of Rome. He became attached to Torricelli in Rome and later on he maintained a lively correspondence with his revered teacher on questions in mathematics and physics. We learn from him that although the *Exercitatio Mathematica* was not published until 1666, it originated so far as its basic ideas were concerned as early as 1645. In it, tangents to "higher ellipses and parabolas" $x^p y^q = z^{p+q}$ (with $ax + by + cz = 0$; hence our $W$-

curves) were determined by a rigorous inequality. An algebraic manuscript of a somewhat later date is still unedited.

Finally, we should not omit to mention the Venetian, Stefano Degli Angeli (1623–1697), who studied with Cavalieri in Rome and worked with Ricci and Sluse. From 1658 on, Angeli published many rather small treatises on the infinitesimal mathematics of special classes of curves (parabolas, spirals, cycloids, etc.). He showed thereby that he was thoroughly familiar with the fundamental ideas of Torricelli's last works even though he lacked the power of agreeable presentation. In 1663, shortly before his call to Padua, Angeli undertook to give private instruction to Gregory who stayed with him for a four year period. Gregory received perfect insight into the infinitesimal works of the Galileist school through his teacher. He developed the material he assimilated extensively and in novel ways.

However, the association between the French and Italian mathematicians which had only recently been strengthened was shortly to be broken off again upon the death of Torricelli, of Cavalieri (1647) and of Mersenne (1648). The blame for this lies above all, with Roberval who made use of inconsequential arithmetic errors in Torricelli's cubature of cycloidal solids of revolution as the basis of a malicious statement to the effect that Torricelli had achieved his result only by the development of suggestions received from the French.

### 3. Extension and Deepening of Gains Made (1648–1665)

A powerful impetus in the field of geometry came from a practical man. G. Desargues (1591–1661), an architect and military engineer of Lyons, devoted much detailed study to the conic sections of Apollonius. His lectures on geometry (given from

1626) were ridiculed by Parisian architects and artists as *Leçons de ténèbres* and this gave rise to a vehement dispute between them and Desargues as to dogmatical attitudes and tendencies to theorize. The controversy necessitated his return to Lyons and it led to the almost complete rejection of the (less important) perspective studies. What was wholly undeserved was that even the pamphlet on the intersection between a plane and a cone (1639) suffered the same fate. It was written, to be sure, in a strange notation which was adorned by flourishes and taken for the most part from botany; but deep and comprehensive concepts in the theory of conic sections were expressed in this work.

Desargues replaced the rigid figures of the Euclidean method of reasoning by figures which were mobile. He introduced the pencil of rays and the sheaf of planes, at the same time expressly taking into consideration the improper base. He coined the concept of involution, treated the harmonic position and studied the complete quadrangle and quadrilateral. By central projection, he passed from the circle to the cone and correspondingly to the cylinder. In this way, he reduced the theory of conic sections to the theory of the circle. He arrived, thus, at the polar theory of the conic sections and of the sphere. He investigated the one parameter family of conic sections and noted that it generated an involution on every straight line. He was led through a spatial consideration (in a development of an observation in Stevin's work) to the theorem on perspective triangles, which was named after him.

Among the contemporary experts in the field, only Fermat and Descartes were prepared to put to one side the formal defects of the presentation and to acknowledge (1640) the novel type of concept as significant. Later, Huygens, too, concurred in this view (from 1671). The projective method of reasoning did not gain influence of any great magnitude until the appearance of

La Hire, who from 1673 on, presented some of the new methods and results in a form more familiar to the geometers of the time. By this time the studies by Desargues had been forgotten and lost in oblivion. It may be that La Hire had some knowledge of them through his father Laurent La Hire who was a friend of Desargues. The first to state the basic concepts again on the same level of generality was Poncelet (1822).

The first mathematical studies by the precocious Blaise Pascal (1623–1662) were written under the direct influence of Desargues. His father, Étienne Pascal, was a highly placed administrative official, who had in his turn, received thorough mathematical training (the limaçon of Pascal is named after him). Blaise was reared, without ever having gone to school, in the spirit of Montaigne's teachings. This boy, who, according to the father's planned course of study for him, was to be kept from studying mathematics too soon, found his way self-taught far through Euclid, and on the basis of the Desargues pamphlet of 1639, he was led to one of his elegant discoveries, the celebrated theorem on conic sections (first publication 1640, revised 1648). Pascal was the author of a comprehensive work on conic sections (1648) in which results achieved by Apollonius were organically united with the new perceptions by Desargues. We are familiar with the contents of the manuscript, lost since 1676, from a review by Leibniz. Amazingly, Pascal's theorem on conic sections completely escaped notice, and became known only after its rediscovery by Maclaurin (1720).

In the summer of 1654, Pascal was stirred to action by several questions proposed by the Chevalier de Méré and he engaged in a correspondence with Fermat on the subject of the proper partition of a stake in the event that a dice game were broken off before the end of the play. In this way, step by step, Pascal gained insight into the general structure of the interrelationships. These were compiled in the *Triangle Arithmétique* (published 1665). Here the additive and multiplicative rule for the construction of binomial numbers was developed intuitively from the arithmetical triangle; the summation formula $\sum_{k=0}^{n} \binom{r+k}{k} = \binom{n+r+1}{r+1}$ was set up (independently of Fermat), and the theory of combinations was related to binomial coefficients. The step-by-step determination of $\Sigma k^p$ from $(k+1)^{p+1}$ with the application to $\int_{0}^{a} t^p \, dt : a^{p+1}$ came even earlier, from notes which were probably affected by Roberval's influence. From them, too, came the formation of the system of residues for $10^k$ (mod. $p$) and correspondingly for $12^k$ (mod. $p$) for several values of $p$, and tests for division which followed from this.

We learn from the then unprinted dedication of the work that there were studies, now lost, on magic squares (bordering problem), on the determination of conic sections by means of five basic elements (points and tangents) and on the extension of the Apollonian contact problems to circles which cut lines at given angles. The reasoning from $n$ to $n + 1$ which Pascal gathered from Maurolico's arithmetic (1575) should be emphasized as a particularly far-reaching rule of proof. Toward the end of 1654, the printing of the *Triangle Arithmétique* was finished, but nevertheless, it was not issued because Pascal had withdrawn from worldly matters following a visionary personal experience. Retiring to quiet *Port-Royal*, stronghold of Jansenism, he renounced science, too, for a period of several years.

Meanwhile, the many-sided Chr. Huygens (1629–1695) was maturing. As the son of the Netherlands diplomat, Constantijn Huygens, who was a friend of Descartes and Mersenne, he grew up in the midst of thriving foreign relations and at a rather early period in his youth, he had already become the recipient of many scientific stimuli. In 1645–46, Huygens, as a law student, was given private instruction by van Schooten in selected works of ancient mathematicians, becoming familiar also with selections from Vieta's writings, from Descartes's *Géométrie* and from Fermat's papers on extreme values and the tangent problem. A great stir was created by his theoretical refutation of the view generally held since Galileo's time, that the line assumed by a suspended chain (catenary) was a parabola (1646), by means of the principle of the conservation of energy. In 1651, Huygens found the error caused by the faulty application of indivisibles in the quadrature of the circle by Gregorius S. Vincentio (1647; cf. I, p. 122), and he showed that the quadrature of a segment of a conic section followed from the position of its centroid. He rejected the infinitesimal methods of Cavalieri (1635, 1647) and of Torricelli (1644) accepting as valid only the Archimedean indirect method of reasoning, which led to successful results even in the difficult special cases. In the development (1654) of a suggestion found in Vieta (see I, pp. 98, 107), he simplified the quadrature of the circle, approximating the segment of the circle through circumscribed and inscribed segments of a parabola. In 1655, he received the degree of *Dr. Jur.* in Angers. Huygens came to know the leading Parisian mathematicians and scientists personally and he sought a connection with Wallis in order to learn the details of the latter's quadrature of the circle.

When the theologian, J. Wallis (1616–1703) of Kent, had already become a mature man, he improved himself by a study of Harriot (1631) and of Oughtred (1647), and his first independent work, the *Sections of an Angle* (1648, printed 1685) reflects their conceptions. Although it was known that he was a royalist, he nevertheless continued to be unmolested under Cromwell who well understood the value of Wallis's services in deciphering an intercepted secret diplomatic report. In 1649, Wallis became the Savilian professor at Oxford and in 1654 he was granted the degree of *DD*. In 1660, when Charles II was called back from exile to ascend the throne, Wallis was appointed as his chaplain.

In 1650, Wallis learned the infinitesimal methods of Cavalieri and Torricelli. He extended them, beginning in 1652, by arithmetical methods and formal transition from results proved for integral exponents to such as had fractional exponents (incomplete induction). Working along this line, he discovered the general quadrature of higher parabolas and hyperbolas. Then he investigated (in modern terminology) $f(p, q) =$ $1 : \int_0^1 (1 - t^{2/q})^{p/2} \, dt = 1 : \int_0^1 x^p \, dy^q$ assuming $x^2 + y^2 = 1$ for integral, non-negative $p, q$. First, through change of variable, he set up the symmetrical formula $f(p, q) = f(q, p)$, and through integration by parts, the reduction formula $p \cdot f(p, q) =$ $(p + q) \cdot f(p - 2, q)$. Then, with the added assumption $f(0, q) = 1$, he perceived the binomial coefficients $\left( \dfrac{p + q}{2} \right)!$ : $\left( \dfrac{p}{2} \right)!$ $\left( \dfrac{q}{2} \right)!$ in $f(p, q)$ for $p, q$ even. In addition, he found that the following relationships held: $f(p - 2, q) < f(p, q) <$ $f(p + 2, q)$ and $f(p - 2, q) \cdot f(p + 2, q) < f^2(p, q)$. Next, Wallis proceeded to odd values of $p, q$, by means of which he set $f(1, 1) = 1 : \int_0^1 \sqrt{1 - y^2} \, dy = 1 : \int_0^1 x \, dy = \dfrac{4}{\pi}$.

Making use of the inequalities formulated, for $p - 1, p, p + 1$ in place of $p - 2, p, p + 2$, by a diminution of the argument (not proved), Wallis obtained a system of inequalities and this yielded the infinite product value for $4/\pi$ named after him. He handled $\int_0^1 \sqrt{1 + y^2}\, dy$ and $\int_0^1 \sqrt{x \pm x^2}\, dx$ in a similar manner. His patron, Lord W. Brouncker (1620?–1684) of Ireland was greatly interested in mathematics. Brouncker used approximate computation to demonstrate the probability that the infinite product value for $4/\pi$ was correct, and with this, he dispelled the misgivings of van Schooten and Huygens concerning this result. We are indebted to him also for the continued fraction

$$\frac{4}{\pi} = 1 + \frac{1}{2} + \frac{9}{2} + \frac{25}{2} + \ldots$$

During the same period, Wallis prepared a series of manuscripts for publication. Of these, the theory of conic sections in coordinate form (1655), an investigation of the angle of contingence (1656) and a general introduction to mathematics (1657), can be mentioned only in passing. His latest results were collected in the comprehensive *Arithmetica Infinitorum* (1656). He dispatched a copy of the work to Fermat for his critical review. After Mersenne's death, Fermat, greatly engrossed in his profession, had sent only his smaller treatises to his Parisian friends, such as the elimination method (1650) and the restoration of Euclid's porisms (1654) with which Girard had been occupied earlier (before 1617). Fermat immediately recognized the weakness of Wallis's method of proof. He doubted the validity of the product value for $4/\pi$, and he was put out by the fact that Wallis had published the quadrature of the higher parabolas and hyperbolas in inadequate form without even mentioning Fermat.

Matters came to a controversy by letter between Fermat and Wallis (1657–1658) in the course of which Fermat first issued a challenge for the solution of $\Sigma y^3 = x^2$ (for example: $\Sigma 7^3 = 20^2$)

and of $\Sigma x^2 = y^3$. Here $\Sigma a$ means the sum of all the divisors of the positive integer $a$ (including 1 and $a$). Van Schooten, Hudde and Wallis foundered on this problem; Frenicle advanced, heuristically, to solutions in large composite numbers, whereas Fermat (without having expressed himself precisely enough) had intended the problem to be solved in prime numbers. In the latter case—except for $x = 1$, $y = 1$ and the example mentioned above —this problem has no other solution. Then, Fermat issued a challenge for the smallest pair of integral solutions of the equation $x^2 - py^2 = 1$ ($p$ integral, positive, not a square), and explicitly for either the general existence proof or the treatment of the difficult cases $p = 61$, 109, 127. These were solved in 1657 by Frenicle by systematic trials and in 1658 by Brouncker through continued fractions, while Fermat made use of an effective (reconstructed) modification of the cyclic method (cf. I, p. 39). The existence proof given by Wallis for the continued fraction method had gaps in it.

The third of the challenges issued, namely, to find $x$, $y$ in $a^3 + b^3 = x^3 + y^3$, was thought of by Fermat in terms of step-by-step repetition of the Bombelli-Vieta rule (cf. I, p. 95). Frenicle gave integral solutions in small numbers which could be stated by a reduction of the problem to equations of the form $u^2 - pv^2 = q$. In a misunderstanding of the whole state of affairs, Wallis imagined that he had concluded the dispute favorably to himself and in 1658, he allowed a (one-sided) selection of the correspondence to be published. In an anonymous *Replique* Fermat established that Wallis had failed to come up to expectations in all fundamental questions.

Fermat's *Transmutatio et Emendatio Aequationum* (1657–58, printed 1679) originated in the contention over Wallis's results. It contained the first general methods of integration (change of variable, term by term integration and integration by parts, substitution). With their help, Fermat squared the lemniscate $a^2(x^2 - y^2) = x^4$ (substitution $tx = ay$), the *folium of Descartes* $x^3 + y^3 = axy$ (substitution $a^2y = tx^2$) and the *versiera* $(a^2 + x^2)y = a^3$ (substitution $x = av/u, y = u^2/a$). He also indicated how the step-by-step evaluation of $\int_0^a \sqrt{a^2 - x^2}\,dx$ and $\int_0^\infty x^2\,dx/(a^2 + x^2)^n$ could be performed. When he was informed of the proposed new edition of the Descartes *Geometria*, Fermat, sharply critical of the Cartesian *genres,* reduced the solution of higher problems to the application of the simplest possible algebraic curves, and in anticipation of Newtonian lines of thought, he made allusions to plane sections of cones of the third order.

At the same time, Huygens, stimulated by the correspondence between Fermat and Pascal on questions in probability, developed a complete theory of dice-playing (printed 1657) and made his first great discovery in the domain of the infinitesimal. Through the refutation occasioned by an erroneous rectification of a parabola by Th. Hobbes (1656), Huygens showed that by using a system of chords between equidistant parallels to the diameter, which could be transformed into a system of tangents, the arc of a parabola could be stretched out and he reduced the rectification to the quadrature of the hyperbola. In a similar manner, he achieved the complanation of the surface of revolution of the second order (1657). The general quadrature of the higher parabolas and hyperbolas was successfully accomplished through the division of the area between the arc of the curve and the tangent to it at its bounding points into triangles by means of suitable interpolation of tangents at points between the ends.

Huygens proposed the quadrature of the conchoid and the cissoid (1658) as a problem to his correspondents and he was astonished by the correct solution achieved by Wallis through incomplete induction (1659). In his correspondence with van Schooten (particularly from 1650 on) who also reported on works written by his pupils, and with Sluse, above all (from 1657 on), cubic and biquadratic problems, extreme value questions, tangents, points of inflection, areas and centroids for special alegbraic curves were treated. The curves were given for the most part as the geometric loci.

The canon R. Fr. de Sluse (1622–1685) of Liege came of a distinguished Walloon family. From 1638 to 1642 he studied in Löwen; then he studied in Rome, where in 1643, having received the degree of *Dr. Jur. Can.*, he devoted himself completely to the study of languages and natural science. Sluse found a close and congenial friend in Ricci and he took great interest in the latter's mathematical works even after he assumed the duties of the canon of Liege (1653). From 1652 on, he made use of the mechanical tangent rule which probably came to him from Torricelli by way of Ricci. As early as 1655, he determined the subtangent of $t$ of one of the algebraic curves given by $f(x, y) = \Sigma a_{ik} \, x^i \, y^k = 0$ by a method of division, neglecting the higher terms. As in the Fermat extreme value rule, this, in its result, was equivalent to $t \dfrac{\partial f}{\partial x} + y \dfrac{\partial f}{\partial y} = 0$. The pearls of Sluse $x^p (ax + b)^q = y^r$, and thus a special case of the Ricci $W$-curves played an especially important part in his correspondence with Huygens.

Cavalieri's pupil and successor in Bologna, the clergyman P. Mengoli (1625–1686) remained aloof from this circle of scientists. In the *Geometria Speciosa* (about 1648, printed 1659), Mengoli, influenced in his formulation of problems by Cavalieri, Torricelli and Ricci, found $\int_{0}^{1} x^p \, (1 - x)^q \, dx = \dfrac{1}{p + q + 1}$:

$\begin{pmatrix} p + q \\ p \end{pmatrix}$ by the application of inscribed and circumscribed stepped figures. In a further development of investigations by Gregorius a S. Vincentio (1647), he approximated logarithms through subseries of harmonic series and from this he found the alternating series $ln\ 2 = \sum\limits_{k=0}^{\infty} \dfrac{(-1)^k}{k+1}$.

In the *Novae Quadraturae* (1650), the sharpening of the method of indivisibles by arithmetization was continued effectively. Here the existence of bounds of a partial sum appeared as a necessary and sufficient condition for the possibility of finding the sum of an infinite series of positive terms decreasing without limit. The divergence of the harmonic series was demonstrated by means of $\dfrac{1}{3n-1} + \dfrac{1}{3n} + \dfrac{1}{3n+1} > \dfrac{1}{n}$. Going further, Mengoli found the sum of series such as the divided reciprocal figurate numbers whose terms were made up of the products of equal numbers of successive terms of an arithmetical progression. He advanced to $\sum\limits_{k=1}^{\infty} \dfrac{n}{k(n+k)} = \dfrac{1}{1} + \dfrac{1}{2} + ... + \dfrac{1}{n}$, yet he despaired of ever being able to represent $\sum\limits_{k=1}^{\infty} \dfrac{1}{k^p}$ by closed expressions.

Mengoli's achievements became known in England (from 1668) through Gregory. They sank into oblivion again because they were presented in obscure and unwieldy form and because, beyond this, they were considered to have been surpassed by the power series method. At a later period (1672), his exact proof of the validity of Wallis's product value for $4/\pi$, was noteworthy. However, it was not properly valued by his contemporaries who lacked the deeper understanding required for the arithmetical rigor of the method of proof.

Pascal, who had not taken an active part in the controversy between Fermat and Wallis, issued a challenge in the summer of

1658 with the offer of a prize, for solutions to be written out in full, of quadratures, cubatures and the determinations of centroids of figures bounded by cycloids. Huygens and Sluse withdrew after the first results. By skillfully using zig-zag figures, Wren demonstrated the existence of the arc length $\sigma$ of the cycloid, which had not been required by Pascal, and showed that it was double the chord $\sqrt{2\,ax}$ of the generating circle.

Wallis and Lalovera competed for the prize. Working painstakingly on individual investigations, they never went further than partial results and these were rejected by Pascal as unsatisfactory (1658). The refusal was based on motives that were not objective. This applied on the one hand to the Englishman and on the other, to a member of the Jesuit Order so bitterly attacked by Pascal in the struggle over *Port-Royal* (*Lettres Provinciales*, 1657). In the subsequent *Histoire de la Roulette* (1658–59), Pascal, in an incredible fit of blindness to the facts, adopted as his own opinion the malicious imputations made by Roberval against Torricelli. The refutations by Wallis (1659), by Lalovera (1660) and by Dati (Vindication of Torricelli, 1663) were comprehensive in content and factually justified, but their style was unsatisfying and they were of no effect.

Pascal's own results were assembled in a collection of fictitious letters (1659) under the pseudonym of Amos Detonville. Their mathematical content was arrived at by geometrical refinement of the infinitesimal procedures used by the Italians, by Roberval and Wallis, and the combination of these procedures with the methods of the Jesuits (cf. I, pp. 121–24). Starting with the circle $x^2 + y^2 = a^2$, Pascal performed the step by step evaluation of the simplest case of the integral $\int s^m\, x^p\, y^q\, dx$, through skillful continued use of the characteristic triangle enclosed by $dx,\ dy,\ ds$. He interpreted his results geometrically and in this way, he also considered wedges of cycloidal solids of revolution, as well as helicoidal and obliquely cut off parts of circular cylinders (ungulae). The most significant single achievement was the proof of the equality of the arcs of the Archimedean spiral and of the parabola, and the reduction of the arcs of the prolate and curtate cycloids to those of the ellipse. Special emphasis should be given to Pascal's artistry in presentation and proof which carried the reader to entirely new points of view.

During these years, the first sketches on the nature of mathematics and the dialectic within the inquiry were originated. Here, Pascal, above all, set himself against the concept of indivisibles insofar as he believed it to be inadmissible. He pressed hard for clear and distinct definitions and for a lucid method of reasoning. These studies were crystallized in the so-called *Logic of Port-Royal* written by Arnauld and Nicole (1660). Of the proposed edition of Euclid's Elements, revised along the lines of the new way of thinking, merely fragments remain. The manuscript was destroyed by Pascal when he became aware of the superiority of the *New Elements* written by Arnauld (about 1660, printed 1667). The latter work was the first step on the path to reform in introductory instruction in geometry and thus, to the preparatory training for deeper insights into the nature and methods of mathematics.

A few months after the dispatch of the Pascal collection of letters, the first volume of the new edition of the Descartes *Geometria* appeared. The most important new material it contained was van Heuraet's general procedure for rectification. H. van Heuraet (1633–1660?) of Haarlem, belonged to the innermost circle of van Schooten's students. Through allusions received by letter to new results pertaining to the parabola achieved by Huygens, he arrived at his own discovery (and at the general complanation of the surfaces of revolution). In particular, he achieved, the algebraic rectification of the semicubical parabola $ay^2 = x^3$. With the appearance of this result, Aristotle's objection to the possibility of rectification (see I, p. 20) was overcome and Descartes's mistaken opinion on the same subject was refuted. Fermat immediately took up the theme. In an extension of Wren's zig-zag figure method for the cycloid, he presented a general proof of existence. He also simplified Huygens' stretching out of the parabola (1660) (known to him only by hearsay). In a supplement to Lalovera's written defense against Pascal, there was a presentation, also, of theorems on surfaces of revolution generated by arcs of cycloids and on the equality of areas and of arcs in higher spirals and higher parabolas. Although these theorems had already become known from Torricelli's posthumous works, they were, nevertheless, new to Fermat. The theorems were proved in detail by Angeli (1660).

The results found in connection with the controversy over the cycloid were developed in secret by Huygens. He determined the evolutes of the conic sections and of the cycloid and he outlined a general theory of evolutes. By carrying over the approximation method for the quadrature of the circle to the hyperbola, he arrived at very accurate approximations in the calculation of logarithms. He also achieved the quadrature of the logarithmic curve (1661, Academy-lectures 1666–67) and the construction of a formula for barometric height (1662). As a most powerful nobleman, Huygens was in contact with almost all of the European mathematicians of repute. In 1660–61, he became personally acquainted with Pascal in Paris, with Oldenburg in London and with Wallis in Oxford. Through Carcavy as intermediary, he received reports of Fermat's latest works. In 1661, the latter sent him his quadrature of the cissoid by integration involving symmetry and an algebraic study on equations which were made solvable in radical form by the procedure of taking $x = u + v$.

In 1662, Fermat sent him his last great discovery, the proof based on purely geometric considerations of the validity of Snell's law of refraction.

Directly after the Descartes *Geometria* appeared, Sluse's *Mesolabum* was published. Here, by placing the old, pure geometry and the new coordinate geometry methods opposite one another for comparison, it was shown that all cubic problems could be solved by straight edge and compasses, provided that a conic section drawn in a fixed position were given. Before 1659, Sluse stated his tangent rule in the algorithmically more usable form $t \Sigma i a_{ik} x^{i-1} y^k + \Sigma k a_{ik} x^i y^k = 0$ (printed 1673). In the endeavor to simplify the computation in the Fermat tangent rule, Hudde (1656, by letter 1658, printed 1713), Huygens (1662, by letter 1663, Academy-lecture 1667, printed 1693) and Newton (1665, by letter 1672, printed 1712) all arrived, independently of one another, and almost simultaneously at this result.

At this point, reference must be made to the significant contributions by J. Hudde (1628–1704), son of an Amsterdam patrician. After Hudde had assumed a high office in the town council, he could engage in scientific studies only occasionally. Hudde was one of van Schooten's most capable students. In 1656 he wrote a manuscript (now lost) on infinitesimal problems in which the quadrature of the hyperbola was given in the form of a logarithmic series (by letter 1662). Van Schooten's *Exercitationes* (1657) contained the discussion of the parabolic intersection of the plane $x = c$, $y = 0$, $z = c$ with the surface $\left[ \dfrac{x}{a} - \left( \dfrac{y}{b} \right)^2 \right]^2 + \left[ \dfrac{x}{a} - 2 \left( \dfrac{y}{b} \right)^2 \right] \cdot \left( \dfrac{z}{c} \right)^2 = 0$, the determination of the greatest width of the *folium of Descartes* and an (incorrect) determination of a centroid in a radial field. In letters written during 1656, solid problems were treated by means of a fixed hyperbola and a circle, cubic equations, reduced by means of

$x = u + v$, were solved and tangent problems and quadratures of the pearls of Sluse were elegantly completed. The determination of the common factors of two polynomials by elimination (1657) was carried over in the *Geometria* (1659). This work also contained the discovery of the double placing of the polynomial $f(x) \equiv \Sigma a_k x^{n-k}$ equal to zero, from $\Sigma (p + kq) a_k x^{n-k} \equiv (p + nq) f(x) - qx f'(x) = 0$, which was used for extreme values, tangents, points of inflection and irrationalities, and which was of service in the depression of an equation on the basis of known relations among the roots (1658). Very interesting special rules were given here for distinguishing irreducible polynomials and for the decomposition of such polynomials as were reducible. From 1665 on, Hudde corresponded with Huygens on questions of probability and in 1671 he communicated with de Witt on annuities.

The booklets on the cycloid and the sine curve by the Jesuit, H. Fabri (1606?–1688) of Dauphiny, which also appeared in 1659, must be regarded as a (non-essential) contribution by an outsider to the controversy over the cycloid. The *Euclides Restitutus* (1658) by the Neapolitan, G. A. Borelli (1608–1679) is of greater interest. This excellent work, a well thought through revision of Euclid, contained a detailed treatment of the theory of parallels which took as its starting point the locus of all points equally distant from a straight line. Borelli and Torricelli were fellow students under Castelli. In 1658, while he was in a library in Florence, Borelli happened on an Arabic translation of the *Conica* I–VII by Apollonius which, demonstrably, had been there since the time of Ferdinand I. He published the work (1661) in collaboration with the Syrian Maronite, Abraham von Ekschelles, an expert in several languages. Annexed to it was a Latin revision of the Archimedean *Choice Theorem* edited (but not very successfully) by Foster (1659). The translation of

Book V of the *Conica* shows that Viviani's restoration had struck it right on all essential points.

Other editions, translations, reconstructions and revisions of ancient classics also deserve complete approbation. On the other hand, the contemporary works in geometry, arithmetic and algebra were badly executed. It is clear from the latter works that the common level of mathematical knowledge can by no means be gauged by the record made by the leading personalities. The new understandings were confined, on the contrary, to a narrow circle of initiates. These scholars were held in association with one another by loose ties through intermediaries possessing great interest in technical matters but who, nevertheless, were short on technical training. In an era when literary looting was the order of the day, it is small wonder that the esoteric were bent upon withholding their methods and inclined to the use of concise allusions only, rarely disclosing their results.

Upon van Schooten's death (1660) the decadence of the flourishing Leyden school set in. Upon the death of Pascal (1662) and of Fermat (1665), France lost the leadership of the field of mathematics. The High Baroque period was at an end. Great personalities united in associations which were recognized by the state and promoted for the benefit of the state. The most important of these professional organizations were the London R[oyal] S[ociety] (from 1660 on; Brouncker, first president, Oldenburg, secretary) and the Parisian *Ac[ademie des] Sc[iences]* (from 1666 on; Duhamel, secretary). They gave the scientific way of life concrete external forms and afforded more effective protection to literary property. At the same time scientific periodicals were established. Foremost among them were the *P[hilosophical] T[ransactions]* (from 1665 on, Oldenburg, editor) and the *J[ournal des] S[cavans]* (from 1665 on; at the outset, under vacillating and changing leadership). The possibility of quick publication of preliminary notices and re-ports was thus assured and the importance of the correspondence among the scholars slowly diminished.

From this time on, Huygens was generally recognized as Europe's leading authority in mathematics. In 1666, he went to Paris as the head of the division of mathematics, physics and astronomy of the *Ac. Sc.* He devoted himself ever more intensely to mechanics and optics. He became the leading exponent of the tendency of thought which was opposed to the new type of en-deavor in the field of mathematics by younger forces character-ized by a lack of deep-seated interest and a growing scepticism.

CHAPTER TWO

# The Late Baroque Period
## (approximately 1665-1730)

### 1. Discovery of the power series method (1665–1675)

In the Late Baroque period, too, the cross section of knowledge in the domain of mathematics was not as yet very significant. It was restricted for the most part to tricks taught purely by drill methods. In contrast to this, leading personalities, developing the ideas of Vieta and Descartes, searched for comprehensive methods by which individual results achieved up to this time could be unified. Our reference is to their efforts alone. The earlier purely geometric method of presentation was supplanted by superior algebraic methods of investigation. The latter led to concise, lucid and easily manipulated symbolism and technique of calculation, preparing for the discovery of the power series method. This became a mighty tool of the new research. Combining the effects of the special branch of science and the tendencies of critical perception, it found its fulfillment in the calculus, the greatest accomplishment of the Late Baroque period.

We are indebted to the talented J. Gregory (1638–1675) for the first compilation of a large number of individual results in infinitesimals. He was the youngest son of a country parson in Scotland, and a grandnephew of Anderson, well known as the editor of Vieta's writings. Gregory, whose father had become

entangled in the bloody disorders of his fatherland, received
his instruction, first from his mother and then from his brother,
David Gregory. He studied in Aberdeen and he wrote his *Optics*
without detailed knowledge of the modern literature appertaining to the subject, basing his work entirely on mathematical
reasoning. Here he proposed a reflecting telescope as the most
appropriate instrument for astronomical observations. Gregory
won the favor of the Scottish nobleman, Moray, who was well
educated in science. Moray introduced his protégé to the London
*RS*. There Gregory became personally acquainted with Collins
(1625–1683), librarian of the *RS*, an expert arithmetician,
greatly interested in higher mathematics. In 1663, Gregory went
to Italy carrying with him letters of recommendation written by
Moray. Through private instruction given him by Angeli in
Padua (1664–68), he gained an insight into the latest results
achieved by the Torricelli circle.

In 1667, Gregory published his first individual discovery, the
quadrature of the elliptical or hyperbolic sector $\Sigma$ by inscribed
and circumscribed polygons $f_k$, $F_k$. In an affine extension of the
Archimedean method (see I, pp. 25, 26), he took $f_0$ (an inscribed
triangle) and $F_0$ (a circumscribed quadrilateral), and by virtue
of $f_{k+1} = \sqrt{f_k F_k}$, he constructed $\dfrac{2}{F_{k+1}} = \dfrac{1}{f_{k+1}} + \dfrac{1}{F_k}$ a double se-
quence converging (this technical term was used here for the
first time) to $\Sigma$ from both sides. He rationalized the process of
calculation pursuant to $t f_0 = u^2(u + v)$, $t F_0 = v^2(u + v)$,
hence $t f_1 = uv(u + v)$, $t F_1 = uv(v + v) = 2 u v^2$. He approximated $\Sigma$ by linear combinations of the lowest $f_k$, $F_k$ and, by
passing to the polynomials $\Phi(f_0, F_0) = \Phi(f_1, F_1)$, he tried to
demonstrate the "analytical impossibility" of the quadrature of
the circle (i. e. the transcendence of $\pi$).

In the *Geometriae Pars Universalis* of the same period (1668), we have the first collection of the original achievements of Italian and English mathematicians in the infinitesimal domain. Everything in this copious work was proved by the rigorous indirect method of Archimedes. However, the exposition given exclusively by the method of pure geometry, was difficult to understand. Gregory wished to divide mathematics not into algebra and geometry, but rather into general and special groups of theorems. He regarded the carrying of essential properties over from one figure to another as a general operation and as typically algebraic (anticipation of the group theory point of view). He was already aware of the significance of individual transcendental functions in the classification of integration problems. He demonstrated the admissibility of rectifications by an improved modification of Fermat's zig-zag method (unknown to him at that time). This was carried out by the use of the length $n$ of the normals to the curve, in $\frac{dx}{ds} = \frac{y(x)}{n(x)}$, as he calculated the part of the lateral surface of a right cylinder cut by the plane $z = y$, above the curve $y(x)$. Gregory determined the subtangent $t$ from the approximation $\frac{y}{t} \approx \frac{y - \triangle y}{t - \triangle x}$ (development of Ricci's method, 1666), restricting himself, however, to algebraic curves. He successfully carried out a general transition from polar to rectangular coordinates by the transformation of surface invariants or arc invariants, $x = r$, $y = r\varphi$ or $\int_0^\varphi r \, d\varphi$, through which the logarithmic spiral became a logarithmic curve.

Gregory rarely cited. He made no claim to the numerous details he brought forward, details with which we have already become acquainted from Fermat, Roberval, Torricelli, Huygens, Wallis, Angeli, Ricci and other authors (although he succeeded occasionally in attaining excellent improvements in method).

His claim was exclusively to the unifying and unimpeachable method. He considered the rectification of the ellipse and the hyperbola as analytical, but of a higher grade than the quadrature of the circle. In an appendix, where a parabola and a circle were given as intersecting, it was noted that (in modern terms) the algebraic sum of the distances of the points of intersection from the axis of the parabola was equal to zero.

In the spring of 1668, Gregory returned to London. His *Vera Quadratura* was given friendly notices by Brouncker in the *PT* and on the basis of this, he was accepted in the *RS*. News of the impending appearance of a quadrature of the hyperbola by Mercator impelled Brouncker to publish his rational quadrature of the hyperbola in the *PT* (April, 1668). Brouncker's method had been mentioned by Wallis as early as 1657 in the dispute with Fermat. It led, by the summation of $2^n$ strips of equal width into which the area of the hyperbola $\int_0^1 \frac{dt}{1+t}$ was divided, to

$$ln\ 2 - \frac{1}{2} = \left(\frac{1}{3 \cdot 4}\right) + \left(\frac{1}{5 \cdot 6} + \frac{1}{7 \cdot 8}\right) + \left(\frac{1}{9 \cdot 10} + \frac{1}{11 \cdot 12} + \frac{1}{13 \cdot 14} + \frac{1}{15 \cdot 16}\right) + \dots \text{ and } 1 - ln\ 2 = \left(\frac{1}{2 \cdot 3}\right) + \left(\frac{1}{4 \cdot 5} + \frac{1}{6 \cdot 7}\right) + \left(\frac{1}{8 \cdot 9} + \frac{1}{10 \cdot 11} + \frac{1}{12 \cdot 13} + \frac{1}{14 \cdot 15}\right) + \dots .$$

Brouncker also employed the method in a general procedure whereby $ln\ (1 + x) - \frac{x}{1 + x}$ or $x - ln\ (1 + x)$ could be represented in the form of a poorly converging series of fractions.

Brouncker's rational representation of the area of a hyperbola was undoubtedly better than the irrational iteration method originated by Gregory. Yet, it was greatly overshadowed by the logarithmic series $ln\ (1 + x) = \int_0^x \frac{dt}{1 + t} = \sum_{k=0}^{\infty} \frac{(-1)^k\ x^{k+1}}{k + 1}$. This series—had it not already been contained in a still unpublished study on the hyperbola by the Swiss, Souvey, (mentioned 1630)—would have been discovered first by Hudde (1656), and then by Newton (1665). The third (independent) discoverer was N. Mercator (1620–1687) of Holstein. As early as 1660, Mercator had gone to England where he lived in wretched circumstances. In 1666, he gave a precise statement which we would write today as $\int_0^x \frac{dt}{cos\ t} = ln\ tg\left(\frac{\pi}{4} + \frac{x}{2}\right)$, for a conjecture made by H. Bond (1649, mentioned in the work of R. Norwood, 1657).

The first part of Mercator's *Logarithmotechnia* (July, 1668) contained a rational procedure for the computation of logarithms (see I, p. 114) and in connection with this, a structure of articulated approximation formulas for logarithms, which was equivalent to $\sum_{k=0}^{n} (-1)^k \binom{n}{k} log\ (a + kb) < 0\ (a, b > 1)$. The main result, the logarithmic series $\int_0^x ln(1+t)dt = \sum_{k=0}^{\infty} \frac{(-1)^k\ x^{k+2}}{(k+1)(k+2)}$, was brought up in connection with the area of the hyperbola. It arose implicitly from a verbal statement of the formula and it was based on inadequate infinitesimal considerations. Wallis completed the series for $ln\ \frac{1}{1-x}$ (July 1668); Mercator, himself, took over Mengoli's designation of "natural logarithms" and from his series, he determined the natural and the decadic logarithms (using the modulus) for the first prime numbers.

As Huygens was convinced of the possibility of representing $\pi$ algebraically, he took exception to Gregory's *Vera Quadratura* (July 1668). In the course of the vehement dispute that arose, Gregory published a rigorous proof for the Mercator series and the combined statement $\int_{-x}^{+x} \frac{dt}{1+t} = ln\ \frac{1+x}{1-x}$, in the *Exercitationes Geometricae* (Oct. 1668), adding elegant quadratures of the conchoid and the cissoid. He also included $\int_0^\varphi tg\, t\, dt$, $\int_0^\varphi \frac{dt}{\cos t}$ and the approximation formulas for these integrals. Later on, he himself indicated that the formulas were incorrect. Apart from this, Gregory gave very accurate approximations for the elliptical and hyperbolic sector which he developed from linear combinations of polynomials in $f_k$, $F_k$. They show that he represented the general functions $f(x)$ by the interpolation series $f(0) + \binom{x}{1} \triangle f(0) + \binom{x}{2} \triangle^2 f(0) + \dots$ . This development is consistent with the procedure used by Gregory (1657) in extending Mercator's method for extrapolation of a small argument difference in the tables (table of logarithms): $a_n \approx 2a_{n-1} - a_{n-2} \approx 3a_{n-1} - 3a_{n-2} + a_{n-3} \approx 4a_{n-1} - 6a_{n-2} + 4a_{n-3} - a_{n-4}$ etc. In particular, he expanded $y = \frac{\sin nt}{n \sin t}$ by means of $x = 2(1 - \cos t)$ by passing to the differential equation $x(4 - x)\frac{d^2y}{dx^2} + 3(2 - x)\frac{dy}{dx} = (1 - n^2)y$ and successive differentiation, into a series of powers of $x$. Using $\int_0^\varphi \frac{\sin nt\, dn}{\sin t}$, he then obtained series which converged rapidly to the approximation of $\pi$. He also found the binomial expansion of $\left(1 + \frac{x}{a}\right)^{p/q}$ (obvious from posthumous works, reports by letter from 1670 on).

Thus, as early as the end of 1668, Gregory was in possession of the so-called Taylor's series (still without the remainder and proof of convergence). He produced the series in the more difficult cases through the introduction of appropriate rationalizing parameters but the actual process of execution was ponderous and minutely detailed. Gregory taught at St. Andrews from 1669 and at Edinburgh from 1674. He simplified his method of presentation consistently with hints received from Collins concerning Newton's latest results. Years before this, Newton had advanced to interesting series expansions.

I. Newton (1643–1727) was the posthumous son of a tenant farmer in Lincolnshire. He was technically inclined, but wholly unsuited to the agricultural occupation for which he was originally intended. From 1661 on, he studied philosophy at Cambridge under the guidance of Barrow as his tutor. It was not until 1664 that he turned to the mathematical sciences. At that time, his teacher—already well known as a distinguished expert in the ancient classics and as the author of a skillfully abridged revision of Euclid—accepted a position at Cambridge. The newly founded Lucasian chair of mathematics which he occupied had few obligations connected with it and Barrow instituted a course of general introductory lectures (1664–66).

Newton made a complete study of the works of Oughtred (1652), van Schooten (1657), Vieta (1646), Wallis (from 1655) and of the Descartes *Geometria* (1649, 1659–61). The impulse to investigate $\int_0^x (1 - t^2)^{n/2} dt$ came from Wallis's *Arithmetica Infinitorium*. Newton calculated the value of $\sqrt{1 - x^2}$ for n even and $> 0$, interpolated for odd values of $n$, and gave a corresponding treatment for $(1 - x^2)^{n/2}$ also. He established the validity of the expansion of $\sqrt[n]{1 - x^2}$ by formal quadrature and in addition, he obtained the same expansion by a formal carry-over of the method of extraction of a root as applied to numbers to its use with letters. In an extension of the last method, abandoning the dubious interpolation process which appeared in connection with Vieta's numerical method of approximation, Newton constructed a general power series method, after the manner in which it had been proved to be useful in the quadrature of the hyperbola (1664–65). At the same time, he transformed

Fermat's tangent method to serve an algorithmic purpose and employed it in the determination of velocity.

The study on the variation of a function was instigated by Barrow's lectures. Here, differentiation processes were denoted by points; integration was occasionally denoted by enclosure in squares. Newton laid no special emphasis upon the system of notation used, even though he applied his symbolic algorithm from the very beginning, to determinations of extreme values, tangents, centroids and curvatures, to quadratures, cubatures, rectifications and complanations. Over and above this, he advanced to the fundamental concepts of gravitation and the science of color. He also participated in the editing of Barrow's *Optical Lectures* (1666–68) which were intended for publication.

Barrow and Newton followed the appearance of the *Logarithmotechnia*, the *Geometria Pars Universalis* and the second edition of Sluse's *Mesolabum* (Oct. 1668) with great interest. A first rate, thoroughly worked out third part had been added to the *Mesolabum*, dealing with infinitesimal problems in higher spirals, W-curves and other algebraic curves by a shortened version of Ricci's extreme value method (1666). It was just at this time that Barrow gave his *Lectiones Geometricae* (1668–69). Summing up the observations made in the entire body of literature appertaining to infinitesimal problems, he presented a survey of the subject from a purely geometric point of view. The characteristic triangle, the determination of the subtangent by division, neglecting terms of higher degree (equivalent to $t \frac{\partial f}{\partial x} + y \frac{\partial f}{\partial y} = 0$), and the application to the inverse tangent problems (simple differential equations), appeared in this work. The inverse character of the tangent and the quadrature problems was emphasized with special clarity. Unfortunately the style of the presentation was so difficult, that this profound work became accessible only after the discovery of the calculus, and thus, it became accessible—even to remote outsiders—at a date when it had already been surpassed in content. News was received from Collins (at the end of 1668) to the effect that, in the progress of his studies, Mercator had found series for the determination of the sine from the angle and the reverse (inaccessible). This impelled Newton, yielding to Barrow's entreaties in this direction, to collect his results to date in the *Analysis* (1669, printed 1711). This work contained, among other things, the series for *sin x, arc sin x, ln* $(1 + x)$, $e^x$ and the elliptic arc and the reversal of series. In addition, there was a noteworthy attempt at proof making use of geometric series. The treatise was favorably reviewed by Collins (July 1669) who desired it to be offered to the press as soon as possible. The results in it were announced in official letters (from the end of 1669) to interested foreign correspondents of the *RS*.

In the fall of 1669, Barrow was called to London as chaplain to Charles II. In 1670, he received the degree of *DD*. Newton, who had become Barrow's successor at the latter's suggestion, immediately continued the development (1670–73) of the lectures (again on optics) which Barrow had already started. He also participated in the preparation of the Barrow *Lectiones Geometricae* for printing.

By a revision of the wording, Newton combined the tangent rule with the power series method (1670). He disclosed the circular zone series $\int_0^x \sqrt{a^2 - t^2}\, dt$ (Apr. 1670) from material in his notes. This was passed on by Collins in his correspondence from that time on, and it was explained by Gregory (Dec. 1670). At Collins's request, Newton (1670–71) gave the logarithmic approximation solution of the annuity problem $ax^{n+1} + b =$

$(a + b)x^n$, which could be used for computation, and also the solution for the partial series $\sum\limits_{k=1}^{n} \dfrac{a}{b + kc}$ of the harmonic series. In 1674, Gregory treated the annuity problem by iteration. In the *Methodus* of the same period (printed 1736) there were comprehensive methods of quadratures by means of series besides numerous examples of geometrical applications. In particular, differential equations were handled in a manner whereby a new procedure involving undetermined coefficients was added. The reduction of quadratures to areas of circles and hyperbolas was significant. This was equivalent to a complete theory of the algebraic integrals $\int \Re \, (x, y) \, dx$ with $y^2 = a_0 x^2 + a_1 x + a_2$. In addition, there were the kinematic generation of algebraic and transcendental curves and the treatment of problems of rectification and curvature in connection with the theory of evolutes.

In 1672, Newton was accepted in the *RS* and he presented reports on his reflecting telescope and the science of color. However, he became entangled in a bitter controversy with critics inside and outside of his country. He was deterred by this from the publication he had been contemplating of the lectures on optics and of the discoveries in the field of infinitesimal mathematics. Mercator's Latin translation of Kinckhuysen's *Algebra* (1661) for which Newton had provided supplemental notes, likewise remained unprinted. In his lectures, Newton restricted himself to a general introductory course of study, the *Arithmetica Universalis* (1673–84, printed 1707), where, for example (in an extension of Girard's procedures; see I, p. 108) the sums of powers of the roots of an equation were determined step by step from the coefficients.

After the clarification of the rule of construction for the circular zone series, which he had at first vainly tried to formulate through a combination of his interpolation series for $\pi$, Gregory announced a profusion of other series. Among these, were the series for *arc tg x*, 1 : *cos x*, *ln tg x*, *ln* (1 : *cos x*) and $\int_0^a \int_0^b \sqrt{1 - \left(\dfrac{x}{a}\right)^2 - \left(\dfrac{y}{b}\right)^2} \, dx \, dy$. There were also the upper bound $x + \dfrac{x^3}{3} + \dfrac{x^5}{5} + ... + \dfrac{x^{2n-1}}{2n-1-(2n+1)x^2}$ for $ln \sqrt{\dfrac{1+x}{1-x}}$ and an elegant treatment of the difficult Kepler problem (see **I**, p. 119). The first solution of this problem was achieved by Wren (1658) by means of a curtate cycloid in a purely geometrical procedure. An interesting rectification of the logarithmic curve was accomplished by a reduction to the quadrature of the hyperbola and by direct series expansion.

In order to avoid the inconveniences arising from his general procedure for the solution of equations of the $n$th degree (having no 2nd term), Gregory fell back upon an erroneous idea of Dulaurens which many contemporary mathematicians had adopted. Dulaurens, a scientific outsider, had asserted, in 1667, that all intermediate terms of an equation could be removed by purely algebraic operations. Gregory multiplied the equation by a form of degree $n(n-2)$ whose coefficients he hoped so to determine that an equation of the $(n-1)$th degree in $x^n$ would result. He was successful with equations of the third and fourth degrees. When he reached the equation of the fifth degree, however, he foundered on difficulties in computation which he was not yet able to discern. Nevertheless, he did anticipate a great deal of the Galois theory correctly.

A pretty, single result in the domain of elementary number theory was the solution of the so-called problem of six squares originated by Ozanam, a student of Billy. The problem was to determine three numbers $u, v, w$, so that the sum and difference of any two of them would always be a square (1675). Its solution was carried out by means of a rationalizing parameter. Gregory's communications by letter to Collins concerning his original mathematical discoveries remained unpublished and unknown to his contemporaries. An abstract of the results compiled by Collins (1676) has not been printed to this day. Not once did Gregory's nephew, David Gregory (1661–1708), whose *Exercitatio Geometrica* (1684) was based on the papers left by his uncle, grasp the meaning of these memoranda. Gregory's notes became available in printed form for the first time in 1939. It would appear from them that Gregory, working as Newton's colleague of equal rank, independently of the latter and by a method at variance with his, had reached the same results. Accordingly, he must be considered with equal justification as codiscoverer of the power series method.

## 2. *Discovery of the Calculus (1673–1677)*

The greatest mathematical achievement of the Late Baroque period was the discovery of the calculus. This is exclusively to the credit of G. W. Leibniz (1646–1716), son of a Leipzig professor.

Leibniz lost his father in 1652, his mother in 1664. He studied philosophy in Leipzig. Especially interested in mathematical lines of thought, he went to Jena for the summer term of 1663 to attend lectures by E. Weigel. Although Weigel was highly regarded as a teacher, he was insignificant as a scientist. In 1664, as a *Mag. Art.*, Leibniz began the study of law. In 1664–65, he defended two theses in law and in 1666, one in mathematics (*Diss. de Arte Combinatoria*). The mathematical dissertation was written in ignorance of the more modern literature (Pascal) related to the subject. It is witness to the fact that the author's technical training was only of limited extent. At this early date, Leibniz expressed ideas which aimed in the direction of the extension of formal logic in the sense of modern logistics.

Leibniz's application for the doctorate was rejected in Leipzig on threadbare grounds and for this reason he went to Altdorf. There he presented his *Diss. de Casibus Perplexis in Jure,* and received his degree. Nevertheless, he declined the academic career which was offered to him. After a short stay in Nuremberg, where he was in frequent association with the Rosicrucians, Leibniz went to Frankfurt. He won the favor of J. Chr. v. Boineburg, one time minister of the Electorate of Mainz, and in 1670 he entered the service of F. Ph. v. Schönborn, Elector of Mainz, as his legal and political adviser. In this capacity, indefatigable in his literary activity and full of high flown plans (*Consilium Aegyptiacum*), he travelled to Paris to serve as an attaché of a diplomatic negotiator. However, after the sudden death of the Elector and his minister (spring, 1673), he was no longer considered for diplomatic missions on behalf of Mainz. Not until he was in Paris, did the turn of events bring about his serious absorption in mathematics and indeed his preoccupation with infinitesimal questions above all. Under the influence of Hobbes's *Elem. Philos.* I, (1655), Leibniz attempted to recast the Euclidean wording of several axioms. In the transformation of the axiom of the whole and the part, he succeeded, independently, in achieving the summation of number series by a method of differences. Conferring with Huygens (at the end of 1672), Leibniz asserted that he could find the sum of all infinite number series. He was put to the proof with the example of the reciprocal triangular numbers. Having found their sum, he proceeded at once to the summation of the accompanying "harmonic" triangle, which he set up in correspondence with the arithmetical triangle (see I, p. 83).

$$\frac{1}{1}$$

$$\frac{1}{2} \quad \frac{1}{2}$$

$$\frac{1}{3} \quad \frac{1}{6} \quad \frac{1}{3}$$

$$\frac{1}{4} \quad \frac{1}{10} \quad \frac{1}{10} \quad \frac{1}{4}$$

During a brief stay in London (Jan. 24–Feb. 20, 1673) Leibniz was introduced to the *RS* by Oldenburg, with whom he had been corresponding since 1670. He brought with him a model of a calculating machine which he had not yet completed and he made a hasty promise to put it in working order within a short time. Unthinkingly, he spoke highly of the method of series summation as something new, in the presence of Pell, one of the greatest experts in the algebra of the time, and he was referred, notwithstanding his remarks, to Mengoli and to Regnauld's corresponding procedures in Mouton (1670). Leibniz failed in his attempt to achieve the summation of partial series of the harmonic series, the series of reciprocal squares and higher powers. Owing to Oldenburg's intercession, Leibniz was accepted in the *RS* in 1673, but he was regarded by the leading authorities among the members, such as Pell, Collins, Hooke and Newton as a dilletanting beginner without adequate capacity for self criticism and knowledge of the literature. This estimate of Leibniz was true for the years 1673–74. It was mistakenly maintained even later by Newton and his adherents and this sentiment was the basis of the attitude of the English mathematicians in the controversy over priority.

Back in Paris, in renewed meetings with Huygens, Leibniz became aware for the first time of the slight measure of his technical knowledge and of the existence of hiatuses in his concepts of the nature of mathematics. Guided by Huygens's friendly advice, he worked painstakingly through the mathematical treatises which up to that time he had been in the habit of perusing only superficially. Thus, from this time on, he studied the *Horol. Osc.* (1673) which he received as a gift, and also the writings of Pascal (1659), Fabri (1659, 1669), Gregorius a S. Vincentio (1647), Descartes (edition of 1659–61) and Sluse (1668)—not precisely line by line, but with increasing understanding and critical turn of mind. In this way rather small new discoveries came into focus at first, important ones appearing later on.

From Pascal, Leibniz learned how $\int y \, ds$ could be calculated for the circle $y^2 = 2ax - x^2$ by a comparison of the characteristic triangle $dx, dy, ds$ with the normal triangle $y, a - x, a$. In the same way as Huygens did (1657, partial results in *Horol. Osc.* III), he found the complanation of surfaces of revolution by means of generalizations. From Fabri, he took the concept of the moment of a force and the orientation of the axes ($X$ — axis downward, $Y$ — axis to the right). In studying Pascal further, Leibniz found the transmutation theorem (spring, 1673): Taking the origin $O$ on the $X$ — axis, Leibniz represented an arc of $y \, (x)$ in quadrant I by means of applicates set down in parallel coordinates. Then he cut the $Y$ — axis in $T$ by a secant through two points of the arc, $P_1, P_2$ and he cut the parallel to the $X$ — axis through $T$ by the applicates through $P_1, P_2$ in $Q_1, Q_2$. Now the $\triangle OP_1P_2$ was equal to half of the parallelogram between the parallel applicates, extending from $X$ — axis to $Q_1Q_2$. Through infinitesimal geometry, the application of this theorem yielded the proof that the area of the sector $OP_1P_2$ (in modern terms) was $\frac{1}{2} \int_{x_1}^{x_2} t\,dx$. Thus, $t = y - x\frac{dy}{dx}$ was the length of the applicate sides of the associated infinitesimal parallelogram; $t(x)$ was the so-called quadratrix (nomenclature by Lalovera, 1651).

The transmutation theorem was in agreement with the rectangular coordinate method for the calculation of areas by means of $\frac{1}{2} \int r^2 d\varphi = \frac{1}{2} \oint (y dx - x dy)$ employed by Gregory (1668) and by Barrow (1670). However, Leibniz used an affine construction and hence his theorem was more general. It was applied to the circular arc $y = \sqrt{2ax - x^2}$ ($0 \leqq x \leqq a$); it yielded $t : a = x : y = y : (2a - x) = \sqrt{x} : \sqrt{2a - x}$ and led to the versiera $2at^2 = x(a^2 + t^2)$ as the quadratrix. The area of the circular segment between 0 and $x$ was $\frac{1}{2} \int_0^x t dx$, and that of the corresponding circular sector $\frac{1}{2} (ay + \int_0^x t dx) = a \left\{ t - \int_0^t t^2 dt : (a^2 + t^2) \right.$ . The integral remaining was handled further, by series division and by term by term integration, in agreement with Mercator's *Logarithmotechnia* which Leibniz had acquired in London and had just studied completely. With $x = a$, the celebrated alternating series for $\frac{\pi}{4}$ was developed and by means of it Leibniz achieved the "arithmetical" quadrature of the circle. Moreover, by placing the stretched out circular arc $s = a \int_0^x dx : y$ perpendicular to the ordinate $y$ in an upward direction he obtained the ordinate of the cycloid. The quadratrix of the segment of the cycloid was $t = s$; thus $ay - s(a - x)$ was double the area of the cycloid segment from $O$. Consequently, the segment up to the point having abscissa $a$ was rational.

It was still 1673 when these results were sent to Huygens, who then requested Leibniz to furnish a proof for the infinite product value of $\frac{4}{\pi}$ originated by Wallis and also a determination as to whether or not the quadrature of the circle by Gregory (1667) was of binding force. Leibniz toiled in vain. In supplemental, thorough study of other Gregory treatises of 1668, where he met again many results which he had thought were new discoveries of his own, Leibniz obtained valuable suggestions. About this time, too, he mastered the Sluse tangent rule.

After the successful completion of his calculating machine (summer, 1674; principal discovery: stepped roller), Leibniz received hints at first, and later individual results, out of the contributions by Newton and Gregory to the theory of series (1675–76), which, at that time, had not yet been printed. These were received from Oldenburg who relied for his information upon Collins. Leibniz, himself, worked through the earlier algebraic literature and in an independent second discovery (first by Fermat 1661, unprinted at that time), he set up all the equations which could be solved by use of the form $x = \sqrt[n]{u + \sqrt{v}} + \sqrt[n]{u - \sqrt{v}}$. He discovered the identity $\sqrt{1 + \sqrt{-3}} + \sqrt{1 - \sqrt{-3}} = \sqrt{6}$ which his contemporaries considered astonishing, and he surmised that $f(x + iy) + f(x - iy)$ was real. Announcements from London concerning the attempts of the English mathematicians to solve higher equations (Gregory; Pell; Wallis, *Algebra;* Newton, *Arith. Univ.*) led him to the idea of reducing the equation $x^n - a_2 x^{n-2} + a_3 x^{n-3} - a_4 x^{n-4} \ldots \pm a_n = 0$ by means of $x = y_1 + y_2 + \ldots + y_{n-1}$ to the auxiliary equation $y^{n-1} - xy^{n-2} + b_3 y^{n-3} - b_4 y^{n-4} \ldots \pm b_n = 0$. He expressed the power sums $s_p$ of the auxiliary quantities $y_k$ in terms of $x$ and the $b_k$ (probably in development of previous discoveries by Girard; see I, p. 108). The power sum $s_n$ began with $x^n - nb_{n-2} x^{n-2}$. By the formation of identities with the initial equation, rational values of $b_k$ and $s_q$ were found in terms of the coefficients $a_r$ of the initial equation. Leibniz's very guarded communications to Oldenburg concerning this material, despite erroneous procedure contained in it, again anticipated the concepts of group theory.

In May 1675, Count E. W. v. Tschirnhaus (1651–1708) arrived in London. By skillful algorithmic treatment of special problems and allusions to a general method of solving the equation of the $n$th degree, which he was still withholding, he acquired a reputation as an algebraist of significance.

Tschirnhaus came of a wealthy Saxon noble family of Lower Lusatia. Well prepared by a resident tutor, he attended the Gorlitz gymnasium, attracted by its paramount interest in mathematics and natural science. While studying at the University of Leyden (from 1668 on) he came under the influence of the Cartesian school. In 1672, he fought in the Dutch Wars on the side of the Netherlands against the French, returning to Leyden in 1673 to continue his studies. These were pursued self taught for the most part. In Amsterdam he became acquainted with friends of Spinoza, probably with the philosopher himself in 1674, with whom he engaged in friendly correspondence despite their divergent views. After a short stay in his native land, he returned to the Netherlands and with Spinoza's letter of recommendation to Oldenburg, he went to England. He made the acquaintance of Wallis in Oxford and of Collins in London. Pell refused to meet Tschirnhaus in order to avoid the suspicion that he may have employed foreign suggestions. Collins reported to Gregory by letter regarding Tschirnhaus, who desired shortly to submit algebraic procedures of a new type for publication in the *PT*. The contributions by Newton and Barrow to infinitesimal mathematics also became the subject of discussion, but Tschirnhaus placed no special value upon these results since they had reference not to Cartesian exact mathematics but only to approximation mathematics.

In the fall of 1675, Tschirnhaus called on Huygens and Leibniz, carrying letters of introduction from Oldenburg. He threw himself into the study of French and acted as resident tutor to high ranking personalities. He planned to accompany a German prince on his Knight's tour, but this plan failed. From Tschirnhaus, Leibniz received information concerning the new events in the Netherlands and in England. He noted Tschirnhaus's preeminence in the algorithmic field, applied himself with renewed vigor to infinitesimal questions and, by analogy with the symbolic language of algebra, he invented the calculus (end of Oct. 1675). At the end of November he reported on this work and also on attempts to construct an all embracing system of ideograms (*Characteristica Universalis*). Tschirnhaus did not by any means recognize the significance of the pertinent material. He considered his own methods, which were related to Gregorius a S. Vincentio's procedures (See I, pp. 121, 2), more effective, notwithstanding their purely formal and incomplete state. Later (1682–83), he published as independent discoveries, half understood matter coming out of these conferences, intermingled with incorrect assertions added to it. These essays, which appeared mainly in the *Acta Eruditorum* (= *AE*) founded in 1682, were signed merely, D. T. They played an important part in the controversy over priority. Tschirnhaus could, to be sure, reduce the equation $x^3 + px + q = 0$ to the pure equation $y^3 = c$ by virtue of $y = x^2 + vx + w$ and an appropriate choice of $v$ and $w$. He could also reduce the equation $x^4 + px^2 + qx + r = 0$ to $y^4 + cy^2 + e = 0$. However, when the case of the general equation of the 5th degree was reached, his general attack on the problem (1676; printed: *AE* May 1683) was recognized by Leibniz (1676?) as ineffectual.

After the invention of the calculus, Leibniz skillfully compiled the principal results achieved by his contemporaries in the infinitesimal domain. These notes were found among his literary remains. He endeavored to obtain the Ramée professorship which had become vacant upon Roberval's death, but as a foreigner he was undesired, and his application was unsuccessful. Though he may have had no deep inclination to do so, he entered the service of Johann Friedrich, Duke of Hanover, as librarian and legal adviser. While he was still in Paris, Leibniz visited Clersellier, and there, in the company of Tschirnhaus he saw the works left behind by Descartes. Both men also studied Pascal's posthumous *Conica* completely. On the basis of Leibniz's favorable expert opinion, this work may have been accepted by a printer, where, in a mysterious manner, it was lost.

In the spring of 1676, Leibniz informed Oldenburg of Pascal's literary remains and of their significance (lost). In May, Tschirnhaus named Descartes as the great and unsurpassed reformer in the field of mathematics (not printed). Collins issued a long statement in opposition to this view. In the first part of it, the so-called *Historiola* (not printed) he gave detailed abstracts of the contents of Gregory's letters, above all, of his contributions to series theory, and he added to this a reproduction of Newton's tangent method of 1672. Because of its reference to Newton, this English manuscript was sent to Cambridge, and it was altered according to his suggestion. On Oldenburg's expressed wish to this effect, it was decisively shortened (*Abridgement*). Reports of the algebraic achievements by Wallis, Gregory and Pell were annexed to these writings. The packet was intended for Leibniz and Tschirnhaus. Its purpose was to show that the English had by this time gone far beyond Descartes. A copy of Newton's letter of June 23, 1676 to Oldenburg was attached. Not in a position to evade Oldenburg's exhortations, Newton, against his own inclination, presented selected parts of his series theory, with the intent of assuring his priority. Yet he did so without disclosing the fundamental principles. Most of the individual problems touched upon, such as, above all, the processes involved in the binomial series, were connected with the material treated in the *Historiola*. The collection chosen by Newton gave the reader the impression that Gregory, too, had been substantially surpassed by Newton.

The papers were not entrusted to the mails, but were brought to Leibniz by messenger (despatched on Aug. 5) leaving London behind time (Aug. 24). Leibniz replied in haste (Aug. 27) inasmuch as he must have been afraid that he would come under dark suspicion because of the delay. Disappointed by Newton's reservations, he presented the series for $\frac{\pi}{4}$, as indeed he had promised, though this was not on the basis of transmutation but rather on the basis of the rationalization transformation $ay = tx$. Very concise statements concerning the series for *sin x* and $e^x$, which Newton regarded as plagiarizations of his own communications, became clear for the first time from the draft (not printed) of the letter. It showed that Leibniz had obtained the series for $y = e^x - 1$ by means of procedure whereby $y = x + \int_0^{\bullet} y \, dx$, and had obtained the series for $y = cos \, x$ by means of $y = 1 - \int_0^x d\xi \int_0^\xi y \, dx$ —in anticipation of the iteration processes of Cauchy and Picard. Leibniz indicated that his general method was also helpful in the handling of differential equations. He spoke of new results through the calculation of centroids and of a comprehensive principle of physics (conservation of energy).

Tschirnhaus replied on Sept. 1 (mss.) with special algebraic items of little account. He was firmly convinced that all mathematical quantities could be represented without the transition to series, that is, by a finite number of nests of radicals. Oldenburg forwarded a rejoinder from Collins's pen which had reached him on Oct. 10 (formerly erroneously dated 1675). Here, among other things, it was shown (on the basis of a question put to Wallis) that the Leibniz series for $\frac{\pi}{4}$ had been unknown up to that time.

Newton expressed himself for the first time on Nov. 3 (letter to Oldenburg). He described in detail how, at first through interpolation, and later through generalizing the numerical method in algebraic form, he arrived at the expansion of $\sqrt{1-x^2}$ and how, step by step, he extended this procedure. Added to this were allusions concerning the calculation of algebraic functions by means of the "analytical parallelogram." He concealed the tangent and quadrature method behind anagrams made up of many letters which must have been unsolvable because of the employment, among other terms, of new technical words (fluent and fluxion). Yet he submitted the representation of the binomial integral $\int x^p (a + bx^q)^r\, dx$ in series form (going on with details carried out in the *Quadratura Curvarum* of about the same period, which appeared as a supplement to the *Optics* in 1704). He also gave examples in infinitesimal geometry. Newton criticized the poor convergence of the Leibniz series produced by $\int_0^1 dx$ : $(1 + x^2)$ and advised of improvements, among which were those resulting from $\frac{\pi}{4} = arc\, tg\frac{1}{2} + \frac{1}{2} arc\, tg\frac{4}{7} + \frac{1}{2} arc\, tg\frac{1}{8}$. He achieved the reversal of series through step by step raising to higher powers, and he explained the reversal of $ln (1 + x)$. He sharply rejected what seemed to him to be a metaphysical standpoint in Leibniz's letter (*ludus naturae* instead of *hujus naturae*, misreading by Collins in the copy of the Leibniz letter sent to Newton), and he let it be felt that he desired to break off the correspondence.

Leibniz's journey to Hanover was not direct, but by way of London (there for the period Oct. 13–29, 1676) over the Netherlands (visit with Hudde). He deposited an (unimportant) algebraic manuscript with Oldenburg and presented his calculating machine. Now, for the first time, he came to know Collins personally, and he obtained the opportunity of examining the *Historiola*, Newton's *Analysis* and other mathematical treatises by Gregory and Newton. He made abstracts (not printed) which were related to series theory but not to analysis, since in this area Leibniz found nothing that was new any more.

It was not until the end of June 1677 that Newton's second letter, in bad copy, reached Leibniz. He answered it by return mail. Leibniz renewed his inquiry concerning details of the series theory, but speedily discovered for himself what was meant and then arrived independently at a second discovery of the method of undetermined coefficients.

Upon Oldenburg's death (1677), Leibniz's scientific relations with England came to an end. During the following years he searched in vain for a competent and interested colleague with whom, in conversations on the subject, he could formulate and clarify mathematical ideas that would well up and trouble them. So, it came about that slowly other more remote tasks and interests moved into the foreground. Later, but little time remained for technical mathematical investigations which for the most part had been stimulated from abroad. The great collected work, the *Scientia Infiniti*, came to a standstill in a state of disorganized preparation. But in letters, treatises and in unpublished notes, there were valuable ideas which anticipated in embryo, the modern basic concepts of axiomatics, logistics and analysis situs.

In addition to other English mathematicians of lesser importance, Wallis (now 60 years of age) above all, was informed by

Collins of the circumstances surrounding the controversy between Leibniz and Newton. In 1669, Wallis made a survey of the technical details. Though these were no longer complete in extent, he nevertheless recognized that a matter of great consequence was being treated here. He selected extracts from Newton's letters for inclusion in the voluminous manuscript of his *Algebra* (1676, printed in English: 1685, revised edition in Latin: 1693). This work afforded an excellent survey of numerous contributions to algebra by English mathematicians, but it did not do justice as required to non-English authors. Wallis published the full text of both Newton letters with the consent of the author, and with Leibniz's agreement he published extracts of letters from Leibniz to Oldenburg as well as several parts of his own correspondence with Leibniz (1697–99). This reprinting was prejudicial because of misreading and because the abridgements which were undertaken changed the sense (unintentionally). Later on, it was drawn in as evidence against Leibniz in the dispute over priority without recourse on the part of the English to the original works (available to this day for the most part).

## 3. Extension of the New Methods (1677–1695)

Newton deliberately withdrew from public discussions in 1676. This attitude had its origin in his experience with critics who, through lack of understanding, opposed the theory of color on which he had reported to the *RS* in 1672. As a result of the economic crisis which affected the London publishers after the great fire of 1666, the attempts made by Collins to have Newton's *Analysis* of 1699 printed, met with failure. Leibniz, too, had no proper possibility of publication available to him. A treatise on the rational quadrature of the cycloid (*JS* 1678) was marred by numerous misprints. Leibniz's attempt to enter into renewed discussions with Huygens failed. In 1678, the latter, after a two year convalescence in the Netherlands was staying in Paris again and was occupied mainly with questions in mechanics and optics. Leibniz encountered a lack of interest both in the case of the *Characteristica Geometrica* and in the transcendental questions which were connected with algebraic problems such as $x^x - x = 24$ (printed: *JS* July 14, 1692). Huygens took just as little interest in Leibniz's tangent determination and in his inverse tangent method (1679–80). With Tschirnhaus, too, no constructive exchange of ideas came about (1677–81) because the required keen insight in mathematics was wanting in this partner.

The year 1682 brought a sudden reversal. The Leipzig theologian, O. Mencke, gained Leibniz and Tschirnhaus as contributors to the newly established scientific monthly, the *AE*. Tschirnhaus went to Paris, where, on the basis of his social standing and the prospect held out of scientific publications, he became a member of the *Ac. Sc.* He also won Huygens as a friend. In 1681, the latter as a Protestant, had been successful in abating the growing oppression of his coreligionists in France by a decisive return to the Netherlands. In the hope that he could secure a pensioned position in the *Ac. Sc.* through his scientific publications, Tschirnhaus permitted publication in the *AE* (1682–83), prematurely, of essays which were incorrectly worded. In these essays, which were concerned with the solution of higher equations, (Tschirnhaus transformation), with tangent and extreme value determination and with the algebraic integration of algebraic integrals, Leibniz's basic concepts of infinitesimal mathematics were falsely presented without mention of the author's name. Now, Leibniz gave public utterance to several of his basic ideas, in concise summaries, which deliberately concealed more than they pointed out.

In the *AE* of Feb. 1682, Leibniz presented the series for $\frac{\pi}{4}$ with hints about the convergence of an alternating series, about the method of summation of series by means of a scheme of differences and the harmonic triangle. In the *AE* of June 1682, the law of refraction was treated (reference being made to Fermat) as an extreme value problem, with allusions to the method of the differential calculus. Leibniz called attention to the error in fundamental concepts of physics made by Descartes. The *AE* of Oct. 1683 contained a very clear determination of the cash equivalent of a prepayment in the calculation of compound interest. In the *AE* of May 1684, Leibniz, in a friendly way, discussed Tschirnhaus's would-be method for the algebraic quadrature of algebraic curves. The assertion by Tschirnhaus, that the (indefinite) integral of an algebraic function would have to be algebraic if a single definite integral of this function were algebraic, was refuted by the example $y^4 - 6a^2y^2 + 4x^2y^2 + a^4 = 0$, in connection with which $\int_0^a y\,dx$ was achieved through the quadrature by strips of the Hippocrates crescent between a quadrant and a semicircle. No doubt Leibniz knew from Lionne that the quadrature of the entire crescent was rational, and that the quadrature of the partial figures was not algebraic. At first Tschirnhaus did not see into the heart of Leibniz's example at all. Fundamentally he persisted in his mistaken belief, and he tried to weaken the binding force of Leibniz's method of reasoning by referring to an infinity of quadrable parts of the crescent (triangles with two arc sides) which, perhaps, he had discovered independently (it is also in Lionne). The *AE* of Oct. 1684 contained the first communication (marred by misprints) on the dif-

ferential calculus. At the close, it was indicated that the Debeaune problem $\frac{dy}{dx} = \frac{y}{a}$ was solved by means of a logarithmic
curve.

Meanwhile, Newton was occupied with a detailed study of the
motion of the celestial bodies. From 1684 on, in his lectures at
Cambridge, he presented his concept of the theory of gravitation. The manuscript of this profound work was printed in 1687
under the title *Philosophiae Naturalis Principia Mathematica*.
Its results were found by means of the calculus of fluxions; however, they were proved in the style of the Barrow method which
was becoming antiquated. In Book I, lemma 28 of this work,
there is an assertion that there is no oval with a quadrable area.
It is probable that Newton encouraged the Scotsman, J. Craig,
who was one of his auditors, to study thoroughly the essays in the
*AE* treating infinitesimal problems. These had appeared anonymously and were designated in the table of contents only by
initials. Craig guessed that Leibniz was the author of the
Tschirnhaus as well as the Leibniz contributions. He wondered
at the astonishing contradictions in these publications, and he
tried (vainly) to solve the problem under discussion—the quadrature of algebraic curves—in the *Methodus* which appeared in
1685 (the manuscript had been considered by Newton before its
printing). Here Leibniz was sharply attacked. Above all, it was
asserted that Leibniz had taken his infinitesimal methods from
Barrow, without mentioning his mentor. In the same year, Wallis's *Algebra* was published, containing the extracts of Newton's
letters of 1676.

In the *AE* of June 1686, Leibniz energetically took up his defense. He described his studies in Paris in broad outlines. He
had made a thorough study of the Barrow lectures (which he
had obtained in London in 1673) only after he had made the

decisive discoveries (the notes which are still unprinted, confirm this; besides, Barrow's difficult style of presentation would have made the work inaccessible to Leibniz as a beginner). Included here was the proof of the transcendence of $\int \sqrt{a^2 \pm x^2}\, dx$. If these (indefinite) quadratures could be carried out algebraically, then there would be a construction for the division of an angle into $n$ equal parts, or for the determination of the $n$th power, which would hold for all numerical values of $n$. In a subsequent place it was emphasized that the tangent method of the differential calculus applied also to transcendental curves. The summation process $\int$ (this symbol in print for the first time) was the opposite of the differentiation process $d$. The manner in which it could be employed geometrically was shown by the example giving a geometrical consideration of the ordinate of a cycloid generated by a unit circle: $y = \sqrt{2x - x^2} + \int_0^x dx : \sqrt{2x - x^2}$.

In Newton's *Principia,* Leibniz's name was mentioned only in passing. He was to be found, in particular, in Book I, lemma 11, *Scholium* of the original edition. Here, Newton referred to the exchange of letters with Leibniz in 1676 and remarked that the latter had told of his own method after the receipt of the allusions in the anagrams concerning the method of fluxions and that this method differed from Newton's only in nomenclature and notation. Therewith, in indirect language, Newton stamped Leibniz as a second discoverer. Nevertheless, in his opinion, second discoverers had no claim to discoveries already made (observation by Newton, May 29, 1716, concerning a letter of April 9, 1716 from Leibniz to Conti). In the second edition of the *Principia* (1713) edited by Cotes and Newton in collaboration, there were added, also, conceptions of the origin of the fact that (infinitesimal) quantities are of different kinds. This was not moderation, but rather intensification, for Newton believed he, rather than Leibniz, had found a better basis for the concept of the differential quotient in his (still inadequate) theory of the "last ratios", with its transition from finite to infinitesimal quantities. In the third edition which was published (1726) with the cooperation of Pemberton, Leibniz was no longer mentioned. But on the other hand, reference was made to the letter of Dec. 20, 1672, to Collins, which contained Newton's tangent method. The Principia was recognized, of course, by contemporary technical experts as an achievement of significance, but it did not by any means meet with complete approval. Thus, in the *AE* of Feb. 1689, Leibniz attempted to present a "better" (in reality incorrect) derivation of the Kepler laws. Huygens added the *Traité de la Lumière* of 1690 (elaboration of the wave theory of light, conceived as early as 1672) to the *Discours de la Cause de la Pesanteur* (preliminary studies from 1667) which was directed against Newton.

A few weeks before the *Principia*, Tschirnhaus's *Medicina Mentis* appeared. This work was rather widely read in Germany. Its contents were mainly logic and theory of knowledge in which lines of thought due to Descartes and Spinoza were developed. Tschirnhaus wished to show how one could, on the basis of inner experience, advance to new perceptions. He illustrated his procedure by examples in infinitesimal mathematics which were oftentimes at hand. However, he presented material which was inaccurate on the ground of inadmissible generalizations, as, for example, the tangent constructions for the "string curves", which originated in the movement of a point ranging over a simple or repeatedly twisted link polygon (funicular polygon). The Swiss Protestant, N. Fatio de Duillier (1664–1753), immediately raised objections to this; Huygens and Leibniz also censured Tschirnhaus for this error.

Fatio studied first in Genf, and then in Paris (1682–86). In

Paris he studied with Cassini, mainly in the field of astronomy. On the occasion of a visit to the Hague (Sept. 1686) where he came in contact with Huygens, he gained Huygens as a correspondent and a friend. In 1687, he also came into close scientific relations with Newton. In 1688, he became a member of the *RS*. About this time, he asserted, in opposition to Huygens, that he could handle the inverse tangent problem directly, hence, without integration. Under his influence, Huygens maintained an attitude of opposition to the Leibniz calculus. Fatio declared Newton to be the first discoverer of the higher analysis and Leibniz to be the second discoverer, dependent upon Newton (by letter 1691, in print 1697). This remark unleashed the decisive phase of the controversy over priority. Fatio's last mathematical achievement was the treatment of the brachystochrone problem (1699). In 1704, after many years in Genf, he returned to England. Fatio came under the spell of the sectarian prophets of miracles, was thrown into prison and placed in the pillory. He died impoverished and abandoned.

In the *AE* of July 1684, Leibniz had already treated the elastic resistance of a weighted beam. In the *AE* of Jan. 1689, he investigated the question of fall in a resisting medium. The essay in the *AE* of Mar. 1686 was the beginning of a long drawn-out discussion concerning the most appropriate measure of force with the unteachable outsider, the Parisian Abbé Catalan. In the course of this discussion, Leibniz issued a challenge (Sept. 1687) for the determination of the curve along which a body would move, falling downward with constant velocity in a gravitational field (Leibniz isochrone). Huygens gave the solution immediately. Leibniz confirmed it in the *AE* of April 1689, although still without the use of the infinitesimal expedient.

Shortly thereafter, in the *AE* of May 1690, Jacques Bernoulli (1655–1705), a professor of mathematics at Basle, presented an analysis by means of the differential calculus. The technical term, integral, originated by Jean Bernoulli (1667–1748), younger brother of Jacques Bernoulli, was employed in print here for the first time.

Jacques Bernoulli was the oldest son of a well-to-do Basle wholesale merchant and town-councillor. Intended for the study of theology, he secured additional training, self-taught, in elementary works, especially in applied mathematics. He became acquainted with the writings of Descartes and Malebranche, and he constructed (1681, 1682) a theory of comets (incorrect). On a journey to the Netherlands (1681–82) and to London (summer, 1682), Jacques Bernoulli came to know the leading natural scientists of his time. He made a thorough study of the Descartes *Geometria* (1659–61) and of the writings of Wallis (1656, 1659), van Schooten (1657) and Barrow (1659, 1670). He wrote on the gravitation of the ether (1682). He gave a course of lectures and demonstrations in Basle on the mechanics of rigid and fluid bodies. From 1677 on, he made entries in an informative diary of studies and scientific sketches (still not printed). Jacques Bernoulli developed Huygens' lottery studies (printed 1657) into a comprehensive and profound theory (printed from his literary remains, 1713). Its principal theorem, the (first) law of large numbers, originated during the summer of 1689 at the latest.

In 1687, Jacques Bernoulli obtained the position of professor of mathematics in Basle, which had become vacant by death. For years he had been giving his younger brother instruction in the mathematical sciences. The latter had escaped commercial studies but had succeeded in wresting from his father no more than permission to study medicine. Now, at the end of 1687, Jacques Bernoulli turned to Leibniz (who was staying in Italy at the time, engaged in genealogical studies) for information concerning matters which were not clear in the treatises on mechanics in the *AE* of July 1684. However, he received Leibniz's reply (1690) at a time when he, together with Jean Bernoulli, had already independently mastered the infinitesimal methods applicable to questions in physics by intensive study (details unknown) of the pertinent older literature and the new Leibniz essays.

In a critical analysis of Wallis's *Ar. Inf.* (1657), Jacques Bernoulli discovered the method of complete induction once again (*AE* Sept. 1686). Nevertheless, even later, he frequently used formal generalizations without proving them rigorously. He mastered the problem of the eighth degree, namely, the division of a general triangle into four equal parts by means of two straight lines perpendicular to each other (*AE* Nov. 1687; similarly solved by Huygens, also, 1673, known for the first time from his literary remains). In addition, he simplified Descartes's graphical methods for the solution of equations by a skillful selection of auxiliary curves (*AE* June 1688), but without knowledge of the studies by Fermat which proceeded along similar lines (printed 1658, 1679).

The First Series Dissertation (1689) by Jacques Bernoulli gave evidence of a slight knowledge of the literature but of a great abundance of inventiveness. It contained the so-called Bernoulli inequality $(1 + x)^n > 1 + nx$ ($x > 0$; $n$ integral, $> 1$; taken from Barrow 1670 without mentioning his name). It also included the proof of divergence for the harmonic series, the summation formulas for the reciprocal figurate numbers and the expansions for $1 : (1 - x)^2$, $1 : (1 - x)^3$, $1 : (1 - x)^4$, without knowledge, however, of Newton's binomial series (accessible through Wallis 1685). The problem of instantaneous interest ($e$-function) treated in the *AE* of May 1690 was also noteworthy.

At the end of the May 1690 treatise, Jacques Bernoulli posed the problem of the catenary. This had, perhaps, crystallized out of the collaboration with his brother who was already mathematically independent, very skillful formally, quick in perception and full of good ideas. The problem was solved immediately by Leibniz (property of a centroid in integral form, application of the logarithmic function and curve). It was also solved by Huygens (method based on geometrical considerations proceeding from the employment of the tangent angle as a parameter) and by Jean Bernoulli (from $dy : dx = a : s$ with auxiliary geometrical considerations in place of the logarithmic function which he did not know at the time). These solutions were printed one directly after the other in the $AE$ of June 1691.

Jacques Bernoulli's first independent treatises on infinitesimal mathematics were published in the $AE$ of Jan. and June 1691. The first part dealt with the tangents, quadrature and rectification (elliptic integral with indicated argument transformation) of the parabolic spiral $x = a\varphi$, $y = a - r$, $cx = y^2$. There were additional contributions on the curvature and evolute. The second part was concerned with the rectification and the area of a sector of the logarithmic spiral and with the spherical helix. However, neither at this time nor later, did Jacques Bernoulli employ polar coordinates. On the contrary, he always performed his calculation on the basis of geometrical auxiliary considerations using Cartesian coordinates. Since he knew as little of the concept of the logarithmic function as Jean Bernoulli, he represented the logarithmic spiral in differential form. Attached to this was the treatment of the form of an inextensible chain of variable thickness and of an extensible chain of constant thickness. He knew nothing of the appearance of the parabolic spiral in the works of Gregory (1668) and Sluse (1668). The unfortunate and factually unjustified observation that Leibniz's calculus, basically, had issued from Barrow's work (1670) through algorithmic reorganization, was used against Leibniz spitefully in the controversy over priority.

In notes of the same period, Jacques Bernoulli obtained the exponential series from the binomial series by means of incomplete induction (in print: 5th Series Dissertation of 1704; similarly in Halley: *PT* 19, No. 216 for Mar.–May 1695). He also obtained the cosine series (in print: *Histoire et Mém. Ac. Sc. Paris* 1702). However, he had come upon these results in the Leibniz treatises in the *AE* of April 1691 where, in addition, the arc tangent series and the spherical helix were treated (extracts from the still unprinted manuscript of 1676 on the arithmetical quadrature of the circle).

In the meantime, with the acquisition of the degree of licentiate, Jean Bernoulli had finished his study of medicine. At the beginning of 1691, he went to Genf to deepen his knowledge of French. There, he gave instruction in the differential calculus to the engineer, J. Chr. Fatio (1656–1720), older brother of N. Fatio, and he continued his own improvement in the technique of the infinitesimal calculus. Among other things, he confirmed the Leibniz series given in the *AE* of April 1691 by means of a rationalizing integral transformation. In the late fall of 1691, Jean Bernoulli travelled back to Paris, where he made the best of impressions in Malebranche's circle which was intensely interested in mathematics. He was considered qualified to give private instruction in higher analysis (lecture notebook received in a later copy and edited) to the Marquis G. Fr. de L'Hospital (1661–1704).

L'Hospital had been intended for a military career, but he had to abandon this field of endeavor because of a high degree of near-sightedness. From this time on, advised in the best possible way by Malebranche who had a sound knowledge of mathematics and who was, also, in contact with Leibniz, he devoted himself completely to the favorite studies of mathematics and mechanics. He corresponded with Huygens from 1690 and he studied Leibniz's first communications on higher analysis. However, it was under Jean Bernoulli's instruction, which was carried on by an exchange of letters (from 1692), that he fathomed the fundamental concepts for the first time. He organized the knowledge he obtained with a skillful hand into a textbook on the differential calculus, which was much used and highly praised in its time. This work (first printing 1696) also contained independent contributions, but the marquis did not give recognition to the part in which Jean Bernoulli had collaborated. Since he had paid a considerable fee for his instruction, he considered that Bernoulli had been paid off and mentioned him only in passing.

In correcting several statements by Tschirnhaus and extending noteworthy general theorems by Leibniz (*AE* of Jan. 1689), the Bernoulli brothers engaged in a friendly competition, studying the basic properties of the catacaustic and diacaustic curves (*AE* of Jan., Mar., May 1692 and June 1693). In the course of this work, Jacques Bernoulli discovered important forms for the representation of the radius of curvature (*theorema aureum*) and, because of their work, Leibniz withdrew his erroneous opinion (*AE* of June 1686) to the effect that the circle of curvature had four consecutive points in common with the curve. Now, Leibniz indicated (*AE* of April 1692) how the envelope of a family of curves could be determined (carried out with a model example: *AE* of July 1694; determination of the coordinates of the center of curvature and the evolute: *AE* of Aug. 1694).

Viviani's problem, to cut four congruent windows out of the surface of a hemisphere so that the remainder of the surface would be quadrable (Florentine problem, pamphlet May 1692) was treated by L'Hospital (lost), Leibniz (*AE* of June 1692), Jacques Bernoulli (*AE* of Aug. 1692), Jean Bernoulli (lost), Huygens (notes of Oct. 27, 1692), Wallis (*PT* 17, No. 197 for Jan. 1693), D. Gregory (*PT* 19, No. 207 for Jan. 1695) and G. Grandi (1699).

As far back as 1638, Debeaune had challenged Descartes to present the treatment of inverse tangent problems leading to logarithmic functions. Descartes discovered the designated tangent properties (letter of Feb. 20, 1639), Leibniz, their connection with the logarithmic curve (notes of July 1676). The controversial treatment of one of these problems, (which led to $\frac{dx}{dy} = \frac{a}{y - x}$) with trivial corollaries found independently by L'Hospital was published by him as his own discovery (*JS* Sept. 1, 1692). Nevertheless, Jean Bernoulli claimed recognition as the author (*AE* of May 1693 and Feb. 1696). On the other hand, L'Hospital seems to have arrived independently (without knowledge of the solution by J. Gregory) at the rectification of the logarithmic curve (summer, 1692), which made a great impression on Huygens and Leibniz. Huygens announced L'Hospital's discovery together with a study on the tractrix (*Hist. Ouvr. Sçavans* Feb. 1693) not suspecting that Jacques Bernoulli had treated this curve just as successfully, as much as two years earlier (not printed). Leibniz brought the Debeaune problem in as an example for the demonstration of the method of undetermined coefficients which he illustrated in the *AE* of April 1693 through the logarithmic series, the exponential series and the cosine series produced from their differential equations. As for the tractrix whose equation he had already set up and solved in Paris, he took up the matter in detail in the *AE* of Sept. 1693. The Bernoulli brothers, generalizing, posed the question of finding the curve through the origin $O$, such that there would be a constant ratio between tangent segment $PT$ up to its intersection with the axis of the abscissas and the intercept $OT$ on the axis of the abscissas (formulation by Jean Bernoulli: $AE$ of May 1693; solution by Jacques Bernoulli: $AE$ of June 1693; generalization by Leibniz: $AE$ of July 1693 and July 1694, and $JS$ of Aug. 23, 1694; solution by L'Hospital: $AE$ of Sept. 1693, extended, $AE$ of May 1694; remarks by Huygens containing the first commendatory mention of the Leibniz differential calculus: $AE$ of Oct. 1693; instrumental solution by Huygens together with the discovery of the vertex: $AE$ of Sept. 1694).

In the Second Series Dissertation (1692), Jacques Bernoulli presented noteworthy series transformations achieved through the additive or, correspondingly, subtractive combining of appropriate terms of unconditionally convergent number series; however, he also took up inadmissible generalizations with respect to conditionally convergent series. The corollaries which were annexed show that just at that time he was working on the classification of curves of the third order (not printed). Meanwhile, Jean Bernoulli had returned to Basle. In 1693, he engaged in a debate on logic. In the spring of 1694 he received the degree of *Dr. Med.* upon the presentation of a celebrated work on muscular movement which was in critical disagreement with Stensen's *Myology* (1667) and Borelli's extensive writing on the movement of animals (1680).

The final composition by Jacques Bernoulli of the great essay on the curve of elasticity was written about this time (printed in the $AE$ of June 1694) after having been announced repeatedly since the first mention of it in the $AE$ of June 1691. Its starting point was the assumption that the elongation of an elastic strip (thought of as having no width) under strain, is proportional to its curvature ($2\rho x = a^2$, hence through a special choice of the

constant of integration: $y = \int\limits_0^x x^2 dx : \sqrt{a^4 - x^4}$, $s = \int\limits_0^x a^2 dx :$
$\sqrt{a^4 - x^4}$). The final result was a general theory in which an
arbitrary function of $x$ took the place of $x$. Exceptional interest
was aroused by the evaluation of the integrals $\int\limits_0^1 dx : \sqrt{1 - x^4}$
and $\int\limits_0^1 x^2 dx : \sqrt{1 - x^4}$ by the employment of the summable comparison series for $\int\limits_0^1 x^3 dx : \sqrt{1 - x^4}$. The results were merely indicated. The derivations were in his notes (printed in part) and
at the end of the Fifth Series Dissertation (1704). In addition,
Jacques Bernoulli treated the problem of the determination of
the curve along which a particle having constant velocity would
move in the earth's gravitational field with respect to a fixed
point (*isochrona paracentria*). This work also appeared in the
*AE* of June 1694. The problem had been repeated many times
since it was first stated by Leibniz in the *AE* of April 1689.
Jacques Bernoulli reduced it to the elliptic integral of the arc of
the curve of elasticity. In the *AE* of Sept. 1694, in place of the
rectification of this curve, he used the rectification of the lemniscate (from the Greek: hippopede) having the equation
$(x^2 + y^2)^2 = a^2(x^2 - y^2)$. Jean Bernoulli made use (independently of Jacques Bernoulli) of this curve (*AE* of Oct. 1694).
That publication also carried Leibniz's own solution through
the rectification of a complicated algebraic curve and Huygens's
indication of the existence of solutions in spiral form.

In May 1694, a memorandum by L'Hospital probably ap-
peared in which, by the example of the cuspidal parabola $a^2 y^3 = x^5$, it was shown, contrary to the statement of Jacques Bernoulli
(*AE* of Mar. 1692) and of Leibniz (*AE* of Sept. 1692), that it
was possible at the point of inflection of a curve, to have $\rho = 0$
beside $\rho = \infty$. Jacques Bernoulli acknowledged his error in the

$AE$ of Sept. 1697. There he explained the circumstances in L'Hospital's example by passing to the limit with $t \to 0$, in the curve $a^2y^3 = x^5 - t^2x^3$ which solved the singularity.

From the end of 1693, Jean Bernoulli engaged in a protracted correspondence with Leibniz. On the basis of an elegant result obtained during the Parisian days, Bernoulli knew of an adroit manner of introducing exponential calculus (printed: $AE$ of Mar. 1697) which yielded the form of the curve $y = x^x$ and the determination of the series for $\int_0^1 y\,dx$. The Bernoulli series $\int_0^x y\,dx = xy - \frac{1}{2!}x^2y' + \frac{1}{3!}x^3y'' \pm \dots$ followed immediately (letter of Sept. 12, 1694; printed: $AE$ of Nov. 1694). Now Leibniz told of results of the Parisian period which were equivalent to Taylor's theorem (letter of Dec. 16, 1694). In the ensuing discussion, Leibniz (1695) developed the main features of a symbolic calculus (restricted in print to differential forms: *Misc. Berol.* 1, 1710), and in the example $y' = y : a$ he advanced to differentiation and integration with fractional indices. A trifling modification through formulas found by successive integration by parts, could have led to the expansion of series with a remainder that could be estimated. This idea did arise in L'Hospital's letter of May 16, 1693 to Jean Bernoulli, but it had no result.

For years there had been a certain tension between the Bernoulli brothers because of the sensitivity of the older one and the surpassing need of the younger one to be esteemed. Malevolent scandalmongers added fuel to the flame. Well meaning friends, like L'Hospital and Leibniz, tried to effect a compromise. Tactless remarks by Jean Bernoulli were printed, and in the $AE$ of Dec. 1695, Jacques Bernoulli reacted with sharp words. The latter failed to recognize the significance of the Bernoulli series and he saw in Jean Bernoulli's interpretation of a differential equation as a direction field (*AE* of Nov. 1694) only the reproduction of the Leibniz method (1673, printed *AE* of Aug. 1694) for the approximation of an integral curve through a polygon. At the same time, he issued a challenge for the solution of the differential equation $y' = p(x) \cdot y + q(x) \cdot y^n$ by means of the separation of variables and for the subsequent quadrature. In the *AE* of Mar. 1696, Leibniz indicated the reduction to a linear differential equation; Jean Bernoulli employed the form $y^{1-n} = \eta = uv$ (letter of Sept. 4, 1696 to Leibniz, printed *AE* of Mar. 1697).

While Jacques Bernoulli, compiling his earlier results, prepared valuable supplementary remarks for publication in the new edition of the Descartes *Geometria* (Frankfurt a. M. 1695), Jean Bernoulli emigrated to Croningen to become a professor of mathematics there, in the late fall of 1695. Jean Bernoulli directed Leibniz's attention to the sale of books left by Huygens who had died in 1695, and he conveyed to Leibniz, observations made by Huygens about essays on infinitesimal problems in the *AE*, observations which were not always objectively critical. It was only during the years 1691–93 that Huygens made an earnest effort to make his way into the calculus. No more resulted from this than the elegant quadrature of the *folium of Descartes* $x^3 + y^3 = axy$ (in agreement with Fermat 1658; carried out in print 1679). To the end, the aging Huygens had at his disposal an extraordinary measure of ability for geometrical representations, but he could no longer become accustomed to the strange and still incomplete algorithm.

In point of fact, there were still many fundamental questions relating to the calculus which had not been clarified; it was not by any means a matter of systematic structure. A very small number of audacious pioneers advanced into new territory.

There was no lack of failure, there was no lack of criticism based on misunderstanding in the ranks of the adherents of the old school, and there was no lack of bitter controversy among the ambitious innovators themselves, each of them aspiring to outdo the others. The most intense of these controversies was the so-called struggle over priority between Leibniz and Newton regarding the discovery of the methods of higher analysis.

## 4. Controversies over the Calculus
### (1695 to about 1730)

*Ce que nous connaisons est peu de chose,*
*ce que nous ignorons est immense.*
Laplace on his death bed.

At first, the Leibniz methods encountered universal lack of understanding. Later, they were fought with increasing violence for various motives. Finally they were at the center of a bitter controversy (dispute over priority) with Newton whose adherents declared that Leibniz was dependent upon Newton. The bellicose Jean Bernoulli, above all, came forward on Leibniz's behalf. He did not always, to be sure, fight with unsheathed weapons and particularly in controversy with Jacques Bernoulli over the treatment of isoperimetric problems, he made use of a dubious expedient.

Tschirnhaus rejected the calculus while its concepts were still in the formative stage because he could not understand the symbolism; Huygens, because he no longer had the power to learn anew. Leibniz's objections to the measure of force inappropriately introduced by Descartes (*AE* of Mar. 1686) led to endless controversies with Catalan, Malebranche and Papin. Even on this occasion, Leibniz tried (*Nouv. Rép. Lett.* May 1687) to advance the concept of limiting values by the consideration of continuity. Later (1691) Catalan who only half understood the Leibniz method and could not even apply it properly, averred that this was an extension of the Cartesian method. He was proved wrong in all essentials by L'Hospital but he was unable to comprehend his own error (controversy: *JS* 1692). The Netherlands doctor, Nieuwentijt, spoke (1694–96) of unjustified neglectful practices in the application of the infinitesimal methods. Leibniz retorted with the statement (*AE* of July 1695) that his method served only for mental abbreviation; that everything without exception could be demonstrated by the Archimedean indirect method which was detailed, indeed. Experts in the new calculus accepted with approval the detailed refutation of other objections by Nieuwentijt, written by Jac. Hermann (1701), faithful pupil of Jacques Bernoulli. Jacques Bernoulli had defended his Third Series Dissertation in Basle in 1696 (application of power series to quadratures and rectifications).

With the acceptance of Varignon (1688) and L'Hospital (1693) in the *Ac. Sc.*, most of whose members were closely connected to Cartesianism, questions in higher mechanics and the calculus became much debated subjects in Paris. Mention may be made of the draw-bridge problem originated by Sauveur (locus of a moving weight which maintains in equilibrium a draw-bridge suspended on a chain over a fixed pulley). Solutions were given by L'Hospital (with supplementary material by Jean Bernoulli), by Jacques Bernoulli (*AE* of Feb. 1695) and also by Leibniz (*AE* of April 1695).

The brachystochrone problem (determination of the curve of fastest descent between two fixed points in the earth's gravitational field) proposed by Jean Bernoulli (*AE* of June 1696), opened the investigations concerned with the problem of variation. They were handled under a condition which was sufficient, to be sure, but nevertheless not necessary, namely that the extreme property of the solution curve remain valid for each of its parts. The *AE* of May 1697 carried the solutions by Leibniz (consideration of extreme value) by Jean Bernoulli (path of light in a stratified medium; the direct treatment involving circular arcs between neighboring points appeared for the first time in the *Histoire et Mémoires de L'Ac. Sc. de Paris = HMP* 1718) by Jacques Bernoulli (consideration of extreme value) and by L'Hospital (incorrect auxiliary consideration). Tschirnhaus (whose solution probably was not independent) and Newton gave only their results. We know through L'Hospital of the false attempts at solutions made by Sauveur and La Hire.

Jacques Bernoulli was stirred to the investigation of orthogonal trajectories and families of curves (*AE* of May 1698) by problems in geometric garb proposed by Jean Bernoulli (*JS* Aug. 26, 1697). Among Jacques Bernoulli's studies were problems such as the determination of the locus of all points *P* which bound equal arcs from *O*, on the curves of a family of curves passing through *O*. Leibniz found the differentiation of an integral with respect to a parameter (letter of Aug. 13, 1697 to Jean Bernoulli, not printed at that time). Jean Bernoulli then issued a challenge for the designation of the shortest lines between two fixed points on a given convex surface. Jacques Bernoulli solved the problem for surfaces of revolution (*AE* of May 1698) and in his notes, he anticipated the vital point of Clairaut's theorem (*HMP* 1733). Jean Bernoulli knew the distinguishing geometrical property of the geodesic lines on any convex surface whatsoever (letter of Aug. 26, 1698 to Leibniz; detailed communication to Klingenstierna 1728, printed 1742).

Jacques Bernoulli, who rightly considered the contemptuous remarks printed by Jean Bernoulli as applying to himself issued a challenge to the latter. In connection with the solution of the brachystochrone problem, he was to give the treatment of the "isoperimetric" problem: to make $\int x^m dy$ a maximum through the choice of an appropriate part of the curve $y(x)$, having a given length. Jean Bernoulli solved the problem immediately upon its receipt through a comparison with hydrostatics and he permitted his method to become apparent (*JS* Dec. 11, 1697). The direct treatment with a restriction to terms of the second order (letter of July 15, 1698 to Leibniz; printed *HMP* 1706) was incorrect; and solely on the basis of allusions made to it, Jacques Bernoulli publicly designated it as wrong. (*JS* Feb. 17, 1698). During the early summer of 1700, Jacques Bernoulli gave the completely correct solution in an open letter (math. abstract: *AE* of June 1700) and in his Dissertation of Mar. 1, 1701 (and in the *AE* of May 1701) he gave his method.

Meanwhile, the *Ac. Sc.* was reorganized by royal decree (1699). Its president was the Oratorian, Bignon, its vice-president, L'Hospital. Leibniz, Newton and the Bernoulli brothers were foreign members. Leibniz and L'Hospital tried in vain to effect a reconciliation between the estranged brothers. No one could take an adequate view of the situation, not even Leibniz, who on the basis of a superficial perusal of Jean Bernoulli's solution, had considered it correct. At the outset, Jean Bernoulli wanted to see the binding force of his procedure confirmed by a board consisting of Leibniz, L'Hospital and Newton and he tried vainly to have his counter statements against Jacques Bernoulli printed in the *AE* and *JS*. After some time he perceived his error. It was only by the death of Jacques Bernoulli that he was spared public disgrace. Forced into taking a position by the appearance of Taylor's *Methodus* (1715), in which Jacques Bernoulli's methods were taken over without any reference to him, and by the publication of Hermann's study relating to the subject (*AE* of Jan. 1718), Jean Bernoulli acknowledged his error (*HMP* 1718, as in *AE* of Jan. and Feb. 1718). Nevertheless, while he took a deal of pleasure in cheap ridicule of inconsequential defects he found in Jacques Bernoulli's presentation, he obtained material simplification by means of geometrical auxiliary considerations (contact point for Euler).

In connection with the erroneous assertion by Tschirnhaus (*AE* of Nov. 1695) that one could associate every arc of a parabola with other arcs by means of construction with ruler and compasses, provided that their ratio was originally rational, Jean Bernoulli developed (*AE* of June 1698) the required algebraic transformation equations. He also determined rational parts of the area of the common cycloid (*HMP* 1699, as in *AE* of July 1699). Jacques achieved the same result (*AE* of Sept. 1699). Jean Bernoulli confirmed his brother's procedure (*AE* of June 1700) and challenged him to represent *sin nt* in terms of *sin t*. He himself reproduced Vieta's formulas (see I, p. 100) which were unknown to him and also those relationships which follow from the separation of the real and imaginary parts of $(cos\ t + i \cdot sin\ t)^n$ (*AE* of April 1701). In the *HMP* of 1702, Jacques Bernoulli represented *sin nt* as a function of *sin t* (unaware of Newton as his predecessor) and *cos nt* as a function of *cos t*. Both brothers extended their formulas to non-integral values of *n* without a rigorous foundation for this extension. Hermann also applied himself to the problem of the division of an angle (*AE* of Aug. 1703).

The Berlin *Sozietät der Wissenschaften* was established in 1700 at Leibniz's instigation. Leibniz was its president, Jablonski, its first secretary. Following Leibniz's suggestion, the Bernoulli brothers and Hermann were chosen as corresponding members. Jacques Bernoulli had broken off his correspondence with Leibniz in 1697 because he regarded the latter (wrongly) as a partisan of Jean Bernoulli in the isoperimetric controversy.

He now became convinced that this was an error and hereafter he was always in rather close contact with Leibniz.

In the *AE* of May 1697, Leibniz imprudently wrote that among the living scientists, apart from the Bernoulli brothers, L'Hospital and Newton,—Hudde was probably the only other scholar capable of mastering the brachystochrone problem which was understandable exclusively to experts. N. Fatio, who as early as 1687 was desirous of advancing to practical infinitesimal methods independently of Leibniz, treated the brachystochrone problem successfully in 1699. Beside this, he demonstrated Newton's procedure (from the *Principia*, Book II, *sect.* 7, *prop.* 34, *schol.*) for the solution of the oldest of the variation problems, requiring the determination of the solid of revolution of least resistance, moving with constant velocity in the direction of its axis in a resisting fluid medium. In the manuscript which was accepted for printing by the *RS*, he reiterated his highlighting of Newton as the first discoverer of the higher analysis and he designated Leibniz as the second, who was, if possible, dependent upon Newton. Fatio's procedure was simplified and reduced to the solution of the differential equation $y \, \frac{dx}{ds} \cdot \left( \frac{dy}{ds} \right)^3 = a$ by L'Hospital (*HMP* 1699, as in *AE* of Aug. 1699) and by Jean Bernoulli (*AE* of Nov. 1699 and May 1700).

As a member of the *RS*, Leibniz raised objections to granting Fatio approval for printing. Sloane, the secretary of the *RS*, apologized for the improper incident as one which was caused by unintentional negligence. Leibniz rejected Fatio's insinuations in dignified terms (*AE* of May 1700). He made commendatory remarks about De Moivre's treatment of infinitesimal problems by the method of undetermined coefficients (*Philosophical Transactions = PT* May 1698) and he added a new method in number pairs and triples (indices) which had greater formal power. The basic ideas of Hindenburg's combinatorial analysis (principal work: 1796) were anticipated here. The Scotsman, Cheyne, who was in contact with Craig, also made use of the method of undetermined coefficients (1703), but rather unskillfully. De Moivre opposed him in this (1704). De Moivre was the only one of Newton's adherents who was on friendly terms with Jean Bernoulli. The earliest survey of the Newtonian theory of fluxions was written by Harris (1702).

In the Parisian *Ac. Sc.*, too, Leibniz had opponents who grudged him his growing influence in science. They took it amiss when Carré, a student of Varignon's, employed the calculus in his writing on integration problems (1700). The prime movers in this opposition were La Hire and the Abbé Galloys, who, as far back as 1676, had prevented the bestowal of a position as professor of mathematics upon Leibniz. The able algebraist, Rolle, pensioned member of the *Ac. Sc.* since 1685, allowed himself to be used as a front. The objections brought forward (1702–05) were based on great misunderstanding and the inability on the part of Rolle to become familiar with the new symbolic language. The objections were so convincingly refuted by Varignon who had been corresponding with Leibniz concerning the matter since 1701, and by Saurin also, that in 1707, Rolle acknowledged that he had been overcome.

In the *JS* of Aug. 3, 1702 and Feb. 15, 1703, Saurin praised L'Hospital for the treatment of indeterminate expressions in the *Analyse*. Jean Bernoulli, opposing him (tactlessly, a few months after L'Hospital's death) stated that the rule and the explanatory examples as well as many other individual parts of the *Analyse* originated with him and not with L'Hospital (*AE* of Aug. 1704; first confirmation of his claim after the publication of the letters of 1693–94). Jean Bernoulli's credibility had been put in question by reason of his dispute with Jacques Bernoulli. By these untimely claims, he frivolously forfeited the confidence of the *Ac. Sc.* In the *HMP* of 1716, where he employed the rule for indeterminate expressions for the determination of a tangent to a curve at a multiple point, Saurin scornfully rejected Jean Bernoulli's claims as unjustified.

In the *AE* of May 1702 and Jan. 1703, Leibniz showed how the integration of rational functions could be performed by the application of partial fractions. The basic ideas for this, suggested, perhaps, by Gregory's treatment of $ln \dfrac{a+x}{a-x}$ (*Ex. Geom.* 1668) originated as early as the Parisian days. Leibniz erred in that he denied that the denominators could be decomposed into linear and quadratic factors and regarded $\int dx : (a^4 + x^4)$, which he could not yet reduce to logarithmic and inverse trigonometric functions, as transcendental. He would not allow himself to be persuaded even by the establishment of facts to the contrary by Jean Bernoulli (*HMP* 1702, extracts, *AE* of Jan. 1703). Here, in addition, Jean Bernoulli recognized the connection between the arc tangent and the logarithm with an imaginary argument, just as Leibniz did (intimated, 1676). Connected to this was an improved derivation of the equation for the division of an angle (*AE* of June 1712).

In the Fourth Series Dissertation (Basle 1698) Jacques Bernoulli carried the treatment of problems in integration forward. The Fifth Series Dissertation (Basle 1704) was defended by his nephew Nicolas (I) Bernoulli. By this time, Jacques Bernoulli had become very ill. He now restricted himself to the reproduction of his earlier notes, which appeared in revisions improved as to method. The indication of the transformation of the series for $ln (1 + x)$ by means of $x = y : (1 - y)$ which strengthened the convergence and which was illustrated by the example $x = 1$, was a noteworthy achievement. It was pointed out, in supplementary remarks, that at "extreme points" of a curve, beside $y' = 0$ and $y' = \infty$, $y'$ could also be indeterminate and assume any value whatsoever. A few weeks later, at Basle, J. Chr. Fatio reported on more rapidly converging representation of the Leibniz series for $\frac{\pi}{4}$ by means of $\frac{1}{2} + \frac{1}{2 \cdot 3} + \frac{1}{3 \cdot 5} + \frac{1}{5 \cdot 7} + \frac{4}{5 \cdot 7 \cdot 9} + \frac{4 \cdot 5}{5 \cdot 7 \cdot 9 \cdot 11} + \dots$ . This new series was proved and brought into conformity with the previous transformation (in letters to Leibniz from Hermann and Jacques Bernoulli). The same transformation was employed successfully by Stirling (*PT.* 30, 1719).

In 1689, Newton entered the English Parliament as the representative of the University of Cambridge. In 1696, he was summoned to the royal mint and in 1699, he took over the conduct of the mint. From 1703 on he was president of the *RS*. He exercised no active influence over the continued development of the calculus, although he did have the *Quadr. Curv.* printed in 1704 (see p. 64). In a critique (anonymous: *AE* of Jan. 1705) presented in well contrived and unassailable form, Leibniz emphasized that Newton's suggestions concerning the calculus of fluxions in the *Principia* had appeared only after his own publication of the differential calculus. Hermann, in a paper written in honor of the deceased Jacques Bernoulli (*AE* of Jan. 1706), inserted a reference to the "great discovery of the century, the Leibniz infinitesimal analysis" without making any mention of Newton, whom he had been in the habit of praising in this connection heretofore. Likewise, Leibniz alone, was given special prominence in Reynau's *Analyse Demontrée* (1708) and Manfredi's work on differential equations (1707) was based exclusively on Leibniz's essays.

Newton's friends were shocked by the slight to the great man. One of them, the Scotsman, Keill, in a study on central forces presented before the *RS* in 1708 (printed *PT* 26, 1710), enlarged upon the fact that his result followed with ease from Newton's calculus of fluxions, and that it was this which Leibniz had published in the *AE* under an alteration of names and notation. Leibniz's letter of protest (1711) led to the appointment of a committee (Mar. 1712) made up for the most part of friends of Newton, who, after a brief examination of the question of priority, rendered a decision against Leibniz (April 1712). They attempted to substantiate the decision (*Comm. Epist.* 1712–13, 2nd ed. 1722 with intensifying supplements by Newton himself, 3rd ed. 1725 with a new foreword) by a reprint of letters, extracts and treatises, most of which had been printed previously (from Wallis, 1699 and the 1711 edition of Newton's *Analysis*). The position taken by the committee had a political background. Newton was a member of the Tories. In 1710–14, the Tories were in control and they were opposed to George, Elector of Hanover and Leibniz's sovereign, who by the *Act of Settlement* (1701) had been made successor to the throne of England. Nevertheless, the decision against Leibniz was wrong. Moreover, it was based on an insufficient foundation, wanting in adequate counterclaim upon the original sources, which are still available today. Superficial examination of the evidence, but not ill will, must be imputed to the members of the committee. As the discoverer of the calculus, Leibniz was independent of Newton. From the time of Mahnke's writing (1926, 1932), in particular, this has finally been established.

Leibniz expressed his opinion in stalwart counterstatements ("broadsheet" of July 29, 1713, *Journ. Lit.* Nov.–Dec. 1713) in which he referred to the massive utterances of June 7, 1713 by Jean Bernoulli. The complete exposition (*Historia et Origo Cal-*

*culi Differentialis* 1714) was not printed at that time. Its details have been proved to be exactly to the point by Leibniz's notes (still unprinted). Newton permitted the publication, without mention of his name, of a report of the *Comm. Epist.* (also of the foreword to the second edition of the *Comm. Epist.*) in the *PT* of Jan. 1715. It contained severe charges against Leibniz, disproved today, e.g., the assertion that the latter had seen the contents of Newton's letter of Nov. 3, 1676 (see p. 64) in London and before it had been forwarded to Hanover.

Attempts made by friends such as Chamberlayne, Montmort and Conti at mediation were fruitless (letters 1714–16). It became a matter of accusations, increasingly unrelenting and ill founded on both sides. Immediately after Leibniz's death (Nov. 14, 1716), Newton allowed the correspondence carried on through Conti as intermediary to be printed as a supplement to the *History of Fluxions* which was compiled by Raphson in the sense of the English conception (printed 1715, edition after the death of the author 1716). A more comprehensive collection in the *Recueil* by Des Maizeaux (1720) was intended to be of service to the cause of Leibniz. The critical edition of all the documents bearing on the question is not yet at hand. The discussion of the question of priority was not carried any further in the succeeding decades. The general view was that Newton had been proved right. Not until the 19th and 20th centuries was the exposition in the *Comm. Epist.* recognized as inadequate in its basis and the decision as incorrect.

Upon Jacques Bernoulli's death, Jean Bernoulli became his successor in Basle and the chief protagonist of the Leibniz infinitesimal methods. He was the center of a circle of students filled with enthusiasm for mathematics, submitting to the despotic whims of the master, who, for all his faults, was surpassingly powerful intellectually. After Leibniz's death, Jean Bernoulli took over the defense of his abused friend. Engaging in numerous scientific contests with success, he confirmed the superiority of the Leibniz method of notation, which, as contrasted with the Newtonian method, was planned to better purpose for the infinitesimal processes. The death of Jacques Bernoulli deprived Jean Bernoulli of the one partner with whom he was truly kindred in spirit, who could be critical in competition. Indeed, he now turned principally to the application of infinitesimal methods in the area of physical and technical questions.

In the *AE* of Aug. 1695, Jean Bernoulli had already treated pairs of curves such that the sum or the difference of their arcs could be rectified through circular arcs. In the *AE* of Aug. 1698, he observed that the curve $y(x)$ by virtue of $\xi = xy'^3$, $\eta = \frac{1}{2}(3xy'^2 - \int y'^2 \, dx)$ could be made to correspond to a second curve $\eta(\xi)$ point by point, so that the sum of the arcs of the two curves would be equal to $x\sqrt{(1 + y'^2)^3}$. In the *JS* of Feb. 12, 1703, he issued a challenge for the construction of algebraic curves having arcs equal to a given arc. Craig's attempts (*PT* Jan.–Feb. 1704, as in *AE* of Apr. 1705) were refuted by Jean Bernoulli (*AE* of Aug. 1705). Now, Jean Bernoulli asserted that a point fixed with respect to a rigid curve which slides along a second fixed algebraic curve, parallel to itself and tangent to the second curve, describes an algebraic curve whose arcs will be equal to the sum or the difference of the arcs of the original curves, depending upon the nature of the tangency. He treated the parallel displacement of the ellipse $\frac{x^2}{b^2} + \frac{y^2}{a^2} = 1$ along the fixed ellipse $\frac{x^2}{a^2} + \frac{y^2}{b^2} = 1$ and always tangent to it, as a characteristic example (letter of Jan. 15, 1707 to Leibniz, as in *Miscellanea Berolinensia* [= *MB*] 1, 1710). In that case, the center of the moving ellipse described a convex curve with $0$ as its center and eight symmetrically placed vertices, tangent externally to the circle about $0$ with radius $a + b$, at four vertices and tangent internally to the circle about $0$ with radius $\sqrt{2(a^2 + b^2)}$, at four vertices. In this way, Jean Bernoulli obtained noteworthy approximations to the perimeter of the ellipse.

Beyond this, Jean Bernoulli proposed the problem of isogonal and orthogonal trajectories (letter of Sept. 12, 1694 to Leibniz). It was connected (*AE* of May 1697) to the concepts in Huygens' wave theory and was given as an exercise in an example of this (orthogonal trajectories of logarithmic curves passing through a fixed point and having the same asymptotes). Jacques Bernoulli solved this special problem (*AE* of May 1698). Jean Bernoulli called attention to the differential equation of the general problem of orthogonal trajectories (*AE* of Oct. 1698). Following an arrangement with Jean Bernoulli, Leibniz, at the end of the communication of Dec. 6, 1715 to Conti, challenged the English to give the orthogonal trajectories of all hyperbolas having the same vertices, from a general point of view. Newton permitted the publication of the ideas of the solution, which he had found in a very short time, in the *PT* of Jan.–March 1716, making reference to Jean Bernoulli. The exposition by Nicolas (II) Bernoulli (*AE* of May 1716, June 1718 and May 1720) was detailed and provided with many examples. Hermann gave other formal improvements (*AE* of Aug. 1717, Aug. 1718, Feb. 1719). Taylor denoted the lengths of the arcs of the trajectories sought by means of a differential equation (*PT* Oct.–Dec. 1717) and Stirling (1717) established that the orthogonal trajectories of the family of hyperbolas proposed by Leibniz were not expressible algebraically. Nicolas (I) Bernoulli also offered a contribution (*AE* of June 1719).

The problem of reciprocal orthogonal trajectories was originated by Nicolas (II) Bernoulli (at the end of the 1720 treatise).

It required the determination of a curve within a parallel strip of a plane, which was to be perpendicular to one of the bounding lines. Two mutually orthogonal families of curves were to be generated from this curve by displacing the curve parallel to the bounding lines or through reflection with respect to the mid-parallel of the strip. Pemberton's attempts at a solution (*AE* 1721–23) were picked to pieces by Jean Bernoulli (*AE* 1721–27) and replaced by others which were more elegant.

Attempts by younger English authors to present a unified theory of fluxions began very early.

The work produced by D. Gregory in 1684 on the basis of the literary legacy left by his uncle, J. Gregory—which he did not fully understand—did not go very deeply into the problem. Neither were the writings of Harris (1702), Cheyne (1703), Hayes (1704), Ditton (1706) and Craig (1718) of much importance. As for the (anonymous) commentary by Crousaz on L'Hospital's *Analyse* (1721) and Stone's 1735 translation of the Method of Fluxions (1730) into French, which in the first part presented an English translation of L'Hospital's work—Jean Bernoulli (1742) raised objections to them, pointing out his right of priority.

R. Cotes (1682–1716) was the most gifted mathematically of all Newton's adherents. At Newton's recommendation he became a professor at Cambridge, occupying the Plume chair of astronomy from 1706. In 1709, he was appointed editor of the second edition of Newton's *Principia* (printed 1713). In the *Logometria* (*PT* Jan.–Mar. 1714), Cotes compiled the earlier contributions of English scientists in a well connected reorganized form. The most important works included were those by Halley (*PT* Mar.–May 1695: $ln(1+x) = \lim_{n\to 0} [(1+x)^n - 1] : n)$ and by De Moivre (*PT* May 1698: functional equation for the definition of the logarithm). In critical disagreement with the Leibniz essays on the integration of rational functions, Cotes discovered the theorem named after him (1716, printed posthumously, 1722; Pemberton's proof there also). Through this theorem, the real decomposition of $x^n \pm a^n$ into linear and quadratic factors was interpreted geometrically. Other contributions to the theory of integration proved to be merely skillful continued developments of selected parts of the best contemporary works in infinitesimal mathematics. Having the knowledge of the Cotes theorem, Taylor issued a challenge by letter (1718) to the mathematicians of the Continent, to find the real integral of $x^{kp-1}$ : $(e + fx^p + gx^{2p})$. The challenging letter contained a critical remark about Leibniz's ineptitude in the treatment of $\int dx$ : $(a^4 + x^4)$. Solutions were given by Jean Bernoulli (*AE* of June 1719; supplements 1742), Hermann (*AE* of Aug. 1719) and Nicolas (I) Bernoulli (*AE* of Oct. 1720).

The *HMP* as early as 1710 contained letters from Hermann and Jean Bernoulli on trajectories produced by the application of central forces. In the *AE* of Feb. 1713, Jean Bernoulli sharply criticized the manner in which Newton treated the motion of the pendulum and the projection of a missile in a resisting medium, in the *Principia*. Keill demanded that he give more specific details (Feb. 1718). Now, Jean Bernoulli treated the fundamental properties of the ballistic curve (*AE* of May 1719); the proof followed in the *AE* of May 1721 (supplement: 1742). The best of Jean Bernoulli's contributions to mechanics were contained in the treatises intended for submission to the *Ac. Sc.* in the prize competition of 1727 (impulse law), of 1730 (explanation of the Kepler movements by means of the Descartes vortex theory) and of 1734 (attempt to reconcile the Newtonian and the Cartesian conceptions in astrophysical questions).

W. Jones (1675–1749) also belonged to Newton's circle of

friends. Originally a shop clerk, he found his way into science, self-taught. In 1706, he wrote an introduction to mathematics. Here he presented the calculation of the hundred place value of $\pi$ performed by the astronomer, Machin, on the basis of the decomposition of $\frac{\pi}{4} = 4 \cdot arc\ tg\ \frac{1}{5} - arc\ tg\ \frac{1}{239}$. In 1711, Jones published Newton's *Analysis* together with reprints of the *Quadratura Curvarum*, the *Enumeratio Linearum*, and the *Methodus Differentialis*. Newton's first attempts in this direction dated as far back as 1675. They led (about 1676) to the so-called Newtonian interpolation formula (by Gregory as early as 1668). The application to numerical quadrature was indicated in Newton's *Principia* (Book III, lemma 5) and was carried out fully by Cotes (from 1707, lectures 1709, printed 1722). The best rule for the purpose was given by Simpson (1743).

Newton's method was applied by Br. Taylor (1685–1731), a student of Keill and Machin, to the derivation of the so-called Taylor's series (manuscript 1712; Taylor knew nothing of the occurrence of this series in the work of J. Gregory). The series was given in the *Methodus Incrementorum* (1715) on the basis of a non-rigorous passing to a limit. The Jean Bernoulli series was also presented (without naming the discoverer). The determination of $\frac{dx}{dy}, \frac{d^2x}{dy^2}, \ldots$ from $\frac{dy}{dx}, \frac{d^2y}{dx^2}, \ldots$ and the determination of singular solutions of differential equations were noteworthy accomplishments. The attempt to treat the vibration of a string mathematically was also significant. This attack on the problem was further developed by Jean Bernoulli in the *Comm. Ac. Sc. Petrop.* (= *CP*) 3, 1728.

In a treatise published anonymously by Jean Bernoulli (*AE* of July 1716), whose authorship Jean Bernoulli stubbornly disavowed and which he did not permit to be included in his *Opera,* Taylor was accused of plagiarization, especially in connection with the Bernoulli series. His attempts at justification were rejected by Nicolas (II) Bernoulli (*AE* of July 1720) and by Burckhardt (*AE* of May 1721). Taylor's method of differences was improved in its procedure by Nicole (*HMP* 1717, 1723, 1724, 1727) and it was applied to the summation of series which were produced from series of reciprocal figurate numbers through the process of generalization. In *PT* 30, 1719-20, there were supplementary contributions by Montmort, Taylor and Stirling. Stirling (1730) presented the series $ln\ n! = \sqrt{2\pi} + \frac{2}{2n-1} ln\ \frac{2n+1}{2} - \frac{1}{2n+1} \mathfrak{P} \left[ \frac{1}{(2n+1)^2} \right]$. On the basis of this series and the employment at the same time of the approximation $e^n \cdot n! \approx n^n \sqrt{2\pi n}$ for large values of $n$, which was given in 1730, and the recursion formula $\binom{n}{2} B_2 + \binom{n}{4} B_4 + \dots = \frac{n-2}{2}$ for the Bernoulli numbers given in that very year, De Moivre developed the series expansion $ln\ n! = ln\ \sqrt{2\pi n} + n(ln\ n - 1) + \sum_{k=1}^{\infty} B_{2k} n^{1-2k} : [(2k-1)\ 2k]$.

In Italy, too, the new concepts were received with great interest. As early as 1703, G. Grandi (1671-1742), an important student of Saccheri, wrote a study in the spirit of the Leibniz infinitesimal mathematics. In 1707, G. Manfredi (1681-1761) compiled works on differential equations written by adherents of the Leibniz method, which had been scattered in various periodicals. Very independent work was done by Count G. C. Fagnano (1682-1766) who at the age of 23 had begun to occupy himself with mathematics autodidactically. He took as his starting point Jean Bernoulli's investigation of curves such that the sum or difference of any two arcs could be represented by a line segment ($AE$ of Oct. 1698; simplest example: $a^2y = x^3$). In 1716, Fagnano determined the elliptical, hyperbolic and cycloidal arc pairs having differences which were rectifiable by elementary methods. In 1717, he advanced to special cases of the addition theorem for the elliptic integral by means of their differential equations. Through considerations which followed from Jacques Bernoulli's results in the $AE$ of Sept. 1694, he discovered that the division of the quadrant of the lemniscate could be carried out algebraically (studies from 1718 on).

J. Riccati (1676–1754), one of Angeli's last students, engaged successfully in contests with Hermann (taught in Padua 1707–13), with Nicolas (I) Bernoulli (taught in Padua 1716–19) and with Nicolas (II) Bernoulli (resident tutor in Venice 1720–22). The contests were concerned with the treatment of infinitesimal problems. We are indebted to him, among other things, for investigations of curves which can be determined by means of $\rho(s)$, $\rho(y)$ and $\rho = r$, and also for interesting studies on the reduction of differential equations of the second order (1722–23). It may be that the works on those cases of the Riccati differential equations which are solvable by elementary means through the separation of variables, ($AE$ $Suppl.$ 8, 1723), and which were also determined by Daniel Bernoulli (1724; $AE$ of Nov. 1725) and by the other Bernoullis, were instigated by a treatise in a manuscript on $y' = x^2 + y^2$ by Jacques Bernoulli. Nicolas (I) Bernoulli knew of this manuscript, which was in a form ready for printing, among papers left by Jacques Bernoulli. A treatise on homogeneous differential equations by Jean Bernoulli was of less importance ($CP$ 1, 1726).

The collective development of the infinitesimal methods was influenced in the first quarter of the 18th century by the contraposition of the adherents of Newton and the adherents of Leibniz. Under the leadership of Jean Bernoulli, who found distinguished helpers in Jac. Hermann (1688–1733), in his nephew, Nicolas (I) Bernoulli (1687–1759), in his sons, Nicolas (II) Bernoulli (1695–1726), departed before his time, and Daniel Bernoulli (1700–1782), the English were outplayed in science. In France, P. Varignon (1654–1722; posthumously 1725) and B. Fontenelle (1657–1757; 1728) brought about widespread interest in the new ideas through their individual presentations and the Leibniz conception prevailed throughout Italy. Mathematicians of the Iberian peninsula maintained an attitude of reserve. The Netherlanders followed hesitatingly. In Germany and in the north countries, very few individuals spoke of the new concepts and they scarcely took any part in the advancing development of the subject. Jean Bernoulli's entry into the ranks of those who favored the Cartesian concepts of mechanics, it must be admitted, resulted in the delay of the acceptance of Newtonian celestial mechanics in France until the appearance of Maupertuis (1698–1759) and Voltaire (1694–1778). During a short stay in England before 1730, these men made their decision in favor of the Newtonian conception.

### 5. Other Mathematical Advances
#### (about 1665 to 1730)

#### Infinitesimal Mathematics in Japan
#### (about 1650 to 1770)

In the six decades, approximately, to which the Late Baroque period of mathematics belongs, the ideas of M. Montaigne, Fr. Bacon, R. Descartes and J. Locke for reforms in pedagogy were widely influential. Many attempts were made to achieve a practical reorganization of the introductory instruction in the individual branches of mathematics. In this connection, we should mention, in particular, the aspirations and activities of W. Ratke (1571–1635), of J. Jungius (1587–1657), of the Jansenists at Port Royal (high point about 1650–1670), of J. B. Schupp (1610–1661) and above all, of J. A. Comenius (1592–1670).

Arithmetic was taught in the mother tongue, for the most part. But, apart from this, it was taught by the older drill methods. Appreciation of the necessity for a well connected structure and for a method of teaching which would lead to profound insights, was still lacking. Instead, a large number of iron handed rules appeared, and these, albeit they were thoroughly useful, placed an unnecessarily heavy burden upon memory. It can by no means be denied that several of the arithmetic textbooks, widely used at that time, contained skillful presentations. Among these were the works of Venturoli (from 1663), of Hodder (from 1671), of Barrème (from 1672) and of Matthiesen (from 1680). Clearly, these books were especially satisfying to the prevailing tastes, but the contact points for more profound considerations were missing. This held true even for the works of so highly esteemed a teacher of that time as E. Weigel. Weigel used up his strength dilettanting with matters which were all too trifling, such as his number system of base four (1673). The capable master arithmetician Meissner (1644–1716) of Hamburg, must be mentioned with praise as the founder of a society devoted to the art of computation, whose goal was the improvement of instruction in arithmetic. Even Leibniz did some incidental work in his notes on numerical questions. Indeed, the *HMP* 1703 contained the first publication of his thoughts on the dyadic system, whose significance for infinitesimal questions was clear to him as early as his Parisian days. Notes on dyadic computation were written by Harriot, also, and (independent) references were made to it by van Schooten (1657). Leibniz must have known nothing of the study on the dyadic system (1670) by Bishop Caramuel y Lobkowitz who was active in a variety of literary fields.

The attempt to reorganize instruction in geometry began with Arnauld (printed 1667) who desired to replace the rigid Euclidean formalism with the more mature structure of his developmental course of study. Arnauld laid great stress on simple and direct methods of reasoning. The Jesuits, Gottigniez (from 1669), Fabri (1669), Pardies (from 1671), Dechales (from 1672), Lamy (1685), and later, Malezieu (1715) and Varignon (1731) worked to the same purpose. N. Mercator (1678) wrote a highly noteworthy introduction to a revision of Euclid by an unknown Jesuit author (1666) who likewise endeavored to achieve a shortened and improved presentation. In the introduction, the simplest geometric loci were explained through the use of motion and of physical models for the purpose of illustration. In most schools, to be sure, the older course of study was preferred. This was confined to greatly simplified selections from Euclid, some trigonometry and exercises in construction. The presentation by Le Clerc (1699) was fairly popular. Great numbers of revisions and translations of Euclid into the language of various countries went a little deeper.

In the Catholic countries, where instruction was based primarily upon the fundamental pedagogical principles of the Jesuits, other comprehensive introductory courses in mathematics, including practical and applied mathematics, were common. The texts were such as those by Tacquet (1669), Guarini (1671), Dechales (1674, 1690), Gottigniez (from 1675), Blondel (1683), V. Giordano (1686) and Ozanam (1693). The mathematical dictionaries by Vitali (1668), Moxon (1680), and Ozanam (1690) were another type of book in this field of literature.

The simplest topics in algebra had, by this time, become a fixed part of the subject matter of the secondary schools. Of course, more profound knowledge was given in the universities. The textbooks by Kersey (1673–74), and Prestet (1675, 1689) are worthy of notice as good collections of the courses usually given at that time. Even logarithms became prevalent—especially in the treatment of astronomical problems. There were tables of common logarithms (mostly 8-places) of which we may mention the much used tables of John Newton (1688). Beside these, tables of natural logarithms by Speidell were also available (likewise from 1688). Elvius, of Sweden, stressed the fact that arithmetical exercises could be solved satisfactorily by tables given to only five places. The technique of computation was slowly adapted to formulas using logarithms.

Nor were infinitesimal questions zealously discussed. The circle of those interested in this field was rather small.

Many problems in recreational mathematics, still popular today, were already included in the works of Leurechon-Oughtred (from 1625) and Ozanam (from 1694). The study on dice games added by Huygens to van Schooten's *Exerc. Math.* (1657) was the starting point for more profound investigations by Jacques Bernoulli. The latter knew neither of the work by Pascal related to this subject (printed 1665), nor of the letters exchanged among Fermat, Pascal, Huygens and Hudde. From 1685 on, Jacques Bernoulli invited discussion of individual problems in questions of probability which he proposed. It is possible that the calculation of the chance of making a winning play (printed 1687) ascribed to Spinoza, was not originated by him but by a Netherlander stimulated by the writings of Jacques Bernoulli. In the progress of his studies, Jacques Bernoulli discovered the law of large numbers (about 1689). Leibniz, who hoped to make new contributions to the art of invention through the mathematical theory of games, (*MB* 1, 1710), received letters from Jacques Bernoulli (1703–04) containing the latter's basic ideas on the subject, but he was no longer able to follow them. The uncompleted manuscript the *Ars Conjectandi*, pointing the way in this field, was prepared for the press in 1713 with the collaboration of Nicolas (I) Bernoulli, from the papers left by Jacques Bernoulli. It contained, among other things, the evaluation of $\sum_{k=1}^{n} k^p$ by means of the Bernoulli numbers, which cannot be dated even at this time, and also valuable thoughts on certainty, necessity, chance, moral and mathematical expectation, probability *a priori* and *a posteriori* and chance of winning when players of different degrees of dexterity participate.

Somewhat earlier than this, Montmort had an investigation on games of chance published (1708). An enlarged edition appeared in 1713. At that time it was already under the influence of the Bernoulli *Ars Conj.* whose contents had been shown to Montmort by Nicolas (I) Bernoulli. The latter wrote an application of the *Ars. Conj.* to life rights in 1709. De Moivre, referring to Montmort (*PT* 27, 1711), arrived independently at the concept of compound probability. He developed his ideas in greater detail in 1720 and 1730. The inquiry into the best adjusted determination of recurring payments in annuities and the contributions to the theory of life and property insurance were also treated by the theory of probability. Graunt presented the results of the observations made over a period of many years of English births and deaths (from 1662). Huygens's notes contained procedures for the graphical evaluation of this material (1669). In connection with the Breslau lists of births and deaths for 1687–91, Halley calculated an extrapolating mortality table for the case of a population of constant size (*PT* 17, 1693). Nicolas (I) Bernoulli and De Moivre were also occupied with problems of this kind.

The algebraists exhausted their finest powers in contentions over a goal which was desired but unattainable, namely, the solution of the general algebraic equation in radical form. Nevertheless, they did achieve noteworthy lesser advances.

Dary, an official dealing with standards of weights and measures, a purely practical man, succeeded (1675) in finding the long sought representation of the three real solutions of an equation of the third degree in the *casus irreducibilis*. In Wallis's Algebra (mss. 1676, English edition 1685, enlarged Latin edition 1693) there were references to Newton's approximation methods for the solution of equations (iteration processes, Newton's approximation rule). These references which were constructed scientifically with very great care and worked out in detail, were carried out somewhat further by Raphson (1690). Rolle developed the cascade method (1690) for the separation and the approximation of the roots of an equation and he noted that extraneous roots were introduced through inappropriate elimination (*HMP* 1708–09). Lagny presented irrational approximations for higher radicals (1692). These were refined by Halley (*PT* 18, 1694, likewise in the supplement to Newton's *Arith. Univ.*, printed 1707) by means of Newton's methods of approximation. Newton's procedures for the determination of the number of complex roots in a given equation (*Arith. Univ.*) were developed by Maclaurin (*PT* 34, 1726; 36, 1730) and by Campbell (*PT* 35, 1728).

Huygens's treatment of the planetarium (1691?, printed 1698) contained fundamental material on the expansion of rational numbers into continued fractions. The method by which the continued fraction expansion for $\sqrt{2}$ could be transformed into a series of unit functions was known previously to Tschirn-haus (letter of Apr. 17, 1677 to Leibniz). Beside this, (at the end of 1675), Leibniz knew how the unit fraction series for $\sqrt{2}$ could be obtained by the continued application of the Babylonian approximation rule (see I, p. 6). Rolle determined the continued fraction expansion, step by step, for $\sqrt{a^2 + b}$ (*Mem. Math. Phys.* 3, 1692). The generalizations for higher radicals given here, had already been shown to be wrong by Jean Bernoulli in the lectures for L'Hospital (1692, printed 1742). On this occasion the Rolle expansion for $\sqrt{a^2 + b}$ was transformed into a series by general methods; in the case of $\sqrt{2}$ Jean Bernoulli pointed out its agreement with the binomial expansion. Lagny attempted, through the use of continued fractions (*HMP* 1719), to clarify the approximation to the $\sqrt{3}$ obtained by Archimedes through computations connected with the circle (see I, p. 27). Only a small assortment of the large number of sketches on algebra by Leibniz have been made available in print. The procedures they contained, although fleetingly indicated for the most part, were interesting. As for his contemporaries, one sketch alone, on questions of notation, was at their disposal. (*MB* 1, 1710).

An observation by Grandi (1703, repeated 1710) started the discussion between Leibniz and his correspondents concerning the value $\frac{1}{2}$ as the sum of $1 - 1 + 1 - 1 \ldots$ . This led to printing of a copy of a letter written by Leibniz (*AE, Suppl.* 5, 1712). Leibniz's remark (*AE* of Apr. 1712), stating that negative numbers do not have real logarithms was attacked by Jean Bernoulli (correspondence 1712–13), who wished to set $ln (- x) = ln x$. The whole matter was cleared up for the first time by Euler (*Hist. Mém. Ac. Bln.* [= *HMB*] 5, 1749). Finally, reference should be made to De Moivre's formula which was merely indicated in the *PT* 25, 1707, in the *Misc.* (1730) and in the *PT* 40, 1738. The modern form first appeared in Euler's *Introductio* (1748).

In the fundamentals of number theory, there is no advance to be recorded over Fermat. Ozanam proposed new problems of interest. Yet he had only special elementary procedures at his command (from 1673). The same may be said for Jaquemet (posthumous papers). In 1690, Rolle solved the indeterminate equation $bx - ay = c$ by the application of continued alternating subtraction. The problem of magic squares was exceptionally popular. Frenicle's enumeration of the 880 possible 16-cell squares (printed 1693) was the most important work in this field.

Numerous individual investigations in elementary geometry afforded a transition to the fundamental laws of perspective and coordinate geometry.

In van Schooten's *Exerc. Math.* (1657), containing a great many single studies in geometry, there was a survey of constructions which could be executed by means of a straight edge and an instrument for transferring line segments. This work was carried further by Mohr (1672). Mohr also gave a detailed treatment of construction with compasses alone (1672). G. Ceva presented an elegant derivation of the triangle theorem named after him, using a mechanical method (1678); a proof by pure geometry was originated by Jean Bernoulli (printed 1742). It was at approximately this time that Prince Rupert of the Palatinate may have proposed and solved the problem of passing one of two congruent dice through the other (Report in the work of Wallis, 1693). An ingenious discovery by L'Hospital connected with the controversy over the characterization of algebraic integrals of algebraic functions was published in *HMP* 1701. Here it was shown that there were an infinite number of parts of the crescent between a quadrant and a semicircle, lying between parallels to the axis of symmetry of the crescent, which could be represented algebraically and which were quadrable by elementary means.

From the numerous sketches that have come down to us, it appears that Leibniz engaged in detailed study of the fundamental questions of geometry and their bearing upon infinitesimal mathematics, without, however, arriving at conclusive results which were satisfying to himself. The most valuable of his efforts in this aspect of his work was the article enclosed in a letter of Sept. 18, 1679 to Huygens, in which Leibniz developed the beginnings of a system of ideograms for denoting topological relationships. Apparently, Huygens's attitude of rejection resulted in Leibniz's loss of the desire to pursue these ideas.

Many authors struggled to achieve the construction of a theory of parallels which would be free of all objections. The theory of parallels had been designated by Savile as early as 1621 as the blemish (*naevus*) upon geometry. Noteworthy attempts at clarification were made by V. Giordano (1680), by Wallis (printed 1693, connected to the attempt by At-Tûsis, which he had translated from the latter's Arabic edition of Euclid of 1594) and by Malezieu (1715). The best work in this direction was presented by Saccheri (1733). Fundamentally, Saccheri had already placed elliptical and hyperbolic geometry side by side with Euclidean geometry, but, nevertheless, he discarded the non-Euclidean cases as based on fallacies in the proof. Mathematicians of that day appreciated these ingenious studies just as little as they did the 1697 concepts of logistics.

Noteworthy contributions to the theory of conic sections were made in Newton's lectures on the *Arith. Univ.* (1673–84, printed 1707) and in his lectures on celestial mechanics (1684–87, printed 1687). The textbook-like expositions by La Hire (1673, 1679, 1685), Ozanam (1687), Craig (1693) and L'Hospital (printed 1707) in which the analytical point of view was not consistently realized throughout, were of less importance. A comprehensive survey of the extant applications of algebra to geometry was given by J. Chr. Sturm (1698) and by Guisnée (1705). The equivalence of the coordinates, firmly established in Leibniz's notes as early as 1675, was never once grasped in a completely logical way by Rabuel (1730).

In the writings by Newton mentioned above, there were also methods of generating higher curves by means of instruments. Newton's classification of the curves of the third order (begun 1676, printed 1704) was supplemented by the works of Stirling (1717) and Stone (*PT* 41, 1740). A great many special algebraic and transcendental curves were discovered in the succeeding period. These included curves somewhat like the rhodenea (Grandi, 1728). For additional details in this very extensive field, reference must be made to the work of G. Loria (from 1902).

Space coordinates were employed by Hudde as early as 1657. Leibniz, too, (notes of 1675) and the Bernoullis (e.g., letter of Feb. 6, 1715 from Jean Bernoulli to Leibniz) were familiar with them. Parent, indeed, wrote equations of surfaces (1705). Pitot (*HMP* 1724) represented helices in terms of space coordinates.

Newton's treatment of curves of the third order was based on the fundamental ideas of perspective. These were brought clearly into prominence by Murdoch (1746). The intensity of the general interest in perspective is indicated by compendious collected works such as that by Dechales (1674) and the more detailed single treatises by La Hire (1673), 'sGravesande (1711), Ditton (1712) and Taylor (1716 and often reprinted).

The first beginnings of the historical digests of mathematics, also, occurred in the Late Baroque period. The attempt by Vossius (1650), despite all its deficiencies, was a work to be taken seriously, whereas the historical sections of the courses given by Tacquet (1645) and Dechales (1690) proved to be disappointing. The English, overall, achieved distinguished results in the editorial field. The collection of authors on the Greek Wars compiled by Thevenot (1693) is no longer satisfying to-day, but on the other hand, D. Gregory's edition of Euclid (1703) was unsurpassed for more than 100 years and Halley's edition of Apollonius (1706, 1710) is unrivalled to-day. Halley based this edition of Apollonius upon preliminary work by Bernard and he employed only the wording obtained from the Arabic. Halley's edition of Menelaus (following the Arabic translation) was not actually published until 1758. It was certainly no accident that the appearance of the great Oxford editions of the Greek mathematicians coincided with the activity of the Earl of Shaftesbury (1671–1713) who died at so early an age. Shaftesbury was the inspired founder of the new humanistic school. Reference must also be made, briefly, to Barrow's revisions of ancient mathematicians (1675) which were written in preparation for his lectures. They contributed greatly to raising the level of mathematical understanding in English universities and in the universities on the continent.

The leading Baroque mathematicians were almost all lone individualists who would not be contented with traditional materials of instruction. With undoubting confidence in their own power of thought and capacity for presentation, they mounted new paths. They shortened the old methods of reasoning by the introduction of appropriate operational symbols. They advanced to a great number of new results and they opened the approaches to other achievements which proved to be in an area where they were within reach. The applicability of the new methods produced numerous successors who no longer had fundamentals at heart, primarily, but rather the extension of formal methods. In this way, the groundwork was laid for the transition to an epoch which regarded matters consistent with reason and definite in aim as its highest ideal. By raising the esteem in which their own concepts were held, they tried to free themselves of the bonds of tradition which, in their opinion, had become unbearable. The Late Baroque led into the age of enlightenment.

While operational infinitesimal methods were being discovered and developed in the West, similar procedures appeared almost simultaneously in the Far East. These developments, uninfluenced by the West, had a characteristic stamp.

As early as the 6th century, individual Buddhist monks came to Japan from Korea, bringing Chinese arithmetic texts with them. Under their influence, a Japanese school of arithmetic arose in the 7th century and flourished thereafter, in which skillful use of the abacus (*soroban*) was also taught. Trade relations were established between the Japanese and Portuguese seafarers in 1543, and with the Dutch in 1609. Francis Xavier was active in Japan from 1549. Christianity spread out rapidly. A Catholic university was established in Funai, where the degrees of *Mag. Art.* and *Dr. Theol.* were granted. In 1600, a movement was started which was directed against the Christians and the Chinese Buddhists. By 1638, it had led to the eradication of Christianity and the complete sealing off of all foreign influence. Such trade relations as remained in existence were under the strictest supervision. Lists of Western publications which had been introduced into Japan have come down to us from that period. They show that only a small number of the arithmetic texts and works on trigonometry and astronomy had reached Japan.

Numerical methods for the solution of higher equations were brought back to Japan from China by Môri, around 1600. His pupil, Yoshida, wrote a comprehensive text in arithmetic (1627), a condensed summary of which was used well into the 19th century. Muramatsu determined the value of $\pi$ to eight decimal places by means of numerical calculation in the $2^{13}$-gon (1663) and Murase employed irrational iteration in the numerical solution of equations of the third degree (about 1680).

The talented Samurai, Seki (1642?–1708), began his studies on the theory of equations along the lines of the Chinese prototypes in 1674. He knew the general solution of the indeterminate equation $bx - ay = 1$ and in 1683, he treated systems of linear equations by the determinant type of procedure. By a fractional iteration process, by continued cutting off of the power series which were yielded and proceeding formally with the results obtained from the first terms, Seki found the smaller of the two positive solution of a quadratic equation. This result, which was in the form of a binomial series, was employed in the construction of a series equivalent to the power series expansion of $(arc \sin x)^2$. Investigations such as these were customarily regarded as secrets of temple science, but they seem to have become known at the imperial court as well. In any case, the world traveller, E. Kämpfer, who may have been in Japan for the period 1690–1693, brought back reports of advanced mathematical knowledge on the part of native born scholars.

Seki's pupil, Takebe (1664–1739) was the discoverer of the Japanese continued fraction method. In 1722, he calculated $\pi$ to 42 places and he expanded $x^2$ by a threefold procedure as a function of $1 - \cos x$. Matsunaga (d. 1744), an indirect pupil of Seki, calculated $\pi$ to 51 decimal places by means of the series for $arc \sin \frac{1}{2}$ (1739) and transformed decimal fractions into common fractions (1740). Arima (1714–1783) carried out noteworthy investigations in number theory in 1763. In 1766, he presented the continued fraction approximation to the value of $\pi$ carried out to 12 and 27 places and also an 8 place continued fraction approximation to the value of $\pi^2$. In 1769, he presented, in addition, the summation formulas for the powers of consecutive numbers and for consecutive figurate numbers. He knew an interpolation formula which was the equivalent of the one given by Newton. Native tradition continued unbroken until the opening of Japan (1868).

Only a small circle of those initiated took part in these investigations. They were concerned exclusively with approximation calculations and formal expansions, not with rigorous methods in our sense of the term. There were also ingeniously worked out elementary studies relating to subjects in recreational mathematics, such as magic squares and magic circles, and beyond this, there were systems of inscribed circles and mutually tangent circles inscribed in triangles and quadrilaterals as well as their corresponding figures in space.

# CHAPTER III

# Age of Enlightenment
## (about 1700 to 1790)

### 1. General Remarks; Elementary Questions

The age of enlightenment bore the stamp of personalities whose basic views emerged from the unification of numerous intellectual tendencies. In England, the enlightenment began with J. Locke (1632–1704) who, despite his personal friendship with Boyle and Newton never attained close familiarity with contemporary exact science. Locke's empiricism came into vogue in England after the removal of the Stuarts in the *Glorious Revolution*. The *Essay Concerning Human Understanding* (1690) made a profound impression, and under its influence, Leibniz presented a detailed treatment of the subject in his *Nouveaux Essais sur l'Entendment Humain* (1707, first printing 1765), written in the light of his own views. Whiston, Newton's successor at Cambridge, who was dismissed from his position in 1711, as an "Arian," belonged to the deistic trend of thought. Shaftesbury was an exponent of the aims of moral philosophy, and S. Clarke, too, shared this attitude of thought. In 1715–16, Clarke engaged in an exchange of polemic writing with Leibniz over the foundations of Newtonian mechanics. The Anglican bishop, Berkeley, who was a leading protagonist of the spiritual tendency, raised philosophical objections to the methods of reasoning in the infinitesimal calculus, as early as 1710. In 1734, he started a vehement controversy by upbraiding the champions of infinitesimal mathematics (he meant Halley) with enticement into free-thinking because of the ostensible impossibility of proving the truth of Christian beliefs, while they, themselves, were wanting in adequate rigor in their own special field. Jurin and Walton directed their attack on Berkeley with inept arguments; Robins presented an excellent defense of Newtonian methods (1735). An exposition of the theory of fluxions by Hodgson (1736) and a generally popular introduction by Rowe (1741) were of less importance. The work produced by Maclaurin (in printing from 1737, issued 1742) referring back to Archimedes for the foundations of the infinitesimal processes surpassed them all. The psychological aspect of empiricism was cultivated by the many-sided D. Hume (1711–1776) who among other things, provided a valuable impulse to the writing of historical works in England and who was in close contact with the French philosophers of the enlightenment.

The age of enlightenment in France was ushered in by the skeptic, P. Bayle (1647–1706). Bayle was the co-publisher of the *Nouv. Rép. Lettres* and the author of the celebrated *Dictionnaire Historique et Critique* (from 1695). Through Maupertuis and Voltaire, Newtonian mechanics and with it the whole world of English empiricism became the subject of discussion in the salons and in the learned societies. The leading intellects of France endured with nothing but loathing the seizures by force of the *ancien régime* in a land completely impoverished by constant wars, by frivolity, by the ostentation and maladministration of the court. In ever increasing proportions, they set the desire for a free and independent way of life against the authoritarian conduct of the state and church. They made the results of theoretical and experimental investigations presented by leading scientists in living languages and in easily understood form, the basis of a general view of life that led to materialism, to a withdrawal from the church and to faith in progress. Their spokesmen were the encyclopaedists.

The first shock came as the result of the sale of a medical lexicon translated from the English by Diderot (1746). Then, Diderot, in association with D'Alembert and others, applied himself to a complete reorganization of the *Cyclopaedia* by Chambers (from 1728), which was to present a survey of the totality of the material of knowledge and education. The most capable experts of France contributed to the great *Encyclopédie* (from 1751). The bond unifying all was the striving for the greatest possible clarity and distinctness in the characterization of each individual fact or circumstance. The contributions in mathematics, coming for the most part from D'Alembert and Condorcet, gave an excellent survey of the condition of the knowledge of the time, at a moment, to be sure, when it was in a state of violent development. Numbered among Diderot's occasional contributors was Rousseau, whose novel, *Émile*, 1762, dealing with character development, met with egregious repercussions.

J. E. Montucla (1725–1799) and the encyclopaedists were kindred in spirit. Montucla was the author of a painstakingly detailed history of the quadrature of the circle (1754) and of a history of mathematics which was a distinguished work for its time (1758). The Jesuit, Bossut, publisher of a highly meritorious edition of Pascal's writings (1779), wrote an outline of the history of mathematics for the *Encyclopédie* which was of rather less importance (1784, enlarged 1802, 1810). The faith of the leading personalities of the French Revolution in the progress of the human race was reflected especially clearly in Condorcet's *Esquisse* (1794 and often reprinted).

In Germany, the establishment of the Prussian University of

Halle paved the way for the transition to the age of enlightenment. A powerful drive for the reorganization of university instruction in mathematics issued from this new educational institution. Here, side by side, the jurist, Thomasius, strongly influenced by Locke, and the pietist, Francke, his adversary, founder and for many years (from 1696) director of the celebrated *Pädagogiums*, were employed as teachers. Here, too, Chr. Wolff (1679–1754), at Leibniz's recommendation, taught from 1707 as a mathematician and from 1709 as a philosopher, as well. An eclectic by nature, and a typical philosopher of the age of enlightenment, Wolff was strongly influenced by Leibniz only at the beginning. Through his much used introductory works (1710, 1713) and the lexicon (1716)—all writings of his first Halle period—he became decisively influential over the instruction in mathematics in the German Evangelical universities and later, especially after the suppression of the Jesuit order (1773), over the Catholic universities. Because of his exclusive employment of rational deductive methods and because of his unremitting intercession for freedom of investigation and of teaching, Wolff was hated to an extreme by the adherents of the authoritarian practice of teaching which had prevailed in the German universities hitherto. He was forced to withdraw from Halle in 1723. At Marburg in Hesse, Wolff was accepted in a friendly way. In 1740, upon the accession of Frederick the Great to the throne, he was recalled to Halle. In 1743 he was installed as chancellor of the university, and in 1745, he was raised to the rank of Baron. As a mathematician, Wolff was actually imitative; he was not productive of original works in mathematics.

L. Chr. Sturm acquired great influence over the instruction in the Evangelical secondary schools of Germany from 1707 through his widely used introductory works. As early as 1708, the Evangelical clergyman, Semler, tried to establish a nonclassical school in Halle for the training of new craftsmen (basically a vocational school). The attempt failed after a few years; a repeated attempt likewise failed (1739). In Berlin the *"ökonomisch-mathematische Realschule"* established (1747) by the pastor, Kecker, who knew of the Semler school, did achieve permanence. It was a technical school divided into several departments.

The Hanoverian university, Göttingen, immediately entered into a rivalry with Halle. Göttingen, established in 1734 in accordance with the plans of Baron v. Munchausen, who was greatly influenced by Locke, was connected with a well equipped library. Segner, to whom we are indebted for fine introductory works, taught here at first. He was followed by Kästner in 1753 (1747, 1757–68), author of the widely used and, for its time, the noteworthy work on the elements of mathematics (1753–66). Kästner was also the author of a history of mathematics (1796–1800), a presentation which did, indeed, lack harmony and was affected by the deficiencies of an older work, but which, nevertheless, in its own way, is of great interest even today. The university in Brunswick, too, the *Collegium Carolinum* (1745), was erected in the spirit of the English empiricists. At Bützow, new site of the Rostock university (1760–89), enlightenment likewise prevailed. Karsten, whose school texts (1767–77, 1780) successfully rivalled those named above, taught here. In the greatly admired secondary school of Dessau, the *Philanthropinum*, which was opened in 1774 by Basedow and skillfully directed by him, Rousseau's ideas of education were already taken into account. Basic concepts of mathematics were also taught; indeed, great emphasis was placed upon intuition (Trapp, Busse) and upon the aim of bringing the students to a proper understanding of the subject matter. Modern pedagogical theory of instruction in mathematics in the elementary school began, of course, with Pestalozzi.

It is possible to do no more than touch slightly upon the great numbers of older and newly added arithmetic texts which were in common use during the period of enlightenment. As far as operational methods were concerned, the advances which appeared were trivial. Introductory works by Clausberg (1732), La Caille (1741), Bézout (1764–69 and 1770–72), Bonnycastle (1780) and Lemoine (1790) written for the professional schools of the Late Baroque and the period of enlightenment, for the academies for young noblemen and for the military schools, were thoroughly noteworthy and very popular. Incidentally, reference should be made to the calculating machines invented by Gersten (*PT* 39, 1735), by Hahn (1774) and by J. H. Müller (1783, described 1786).

In schools which were bound to tradition, instruction in elementary geometry followed the Euclidean course of study. It was given for the most part, to be sure, in the language of the country, with appropriate reorganization and shortening of the demonstrations and with explanatory additions. This was the form of presentation prevalent, especially in England, where the distinguished revision by Simson (from 1756) had a great following. On the other hand, in France, Euclid was almost completely excluded from the course of study. Here, Clairaut's introductory work (1741) was highly successful, whereas Bertrand's course (1778) met with comparatively little response. The best of this kind of work was written by Legendre (from 1794). Even here, it must be admitted, procedures were not fully attained which were genetic and at the same time free of objections from the point of view of system, as had been demanded by La Chapelle (1746) and then by D'Alembert (*Enc. Méthod.*).

Out of the great abundance of individual investigations in elementary geometry, only the most important may be given special attention. The studies cn quadrable figures formed by two intersecting circular arcs (Daniel Bernoulli: 1724; enlarged by Cramer: *HMB* 4, 1748; Wallenius: 1761; Euler: *NCP* 16, 1771) were extended to figures between the arcs of two conic sections by young Clairaut (1731). Maupertuis handled the rolling of one regular polygon upon another (*HMP* 1727; then Meister: *Novi Comm. Soc. Göttingen* 1769–70). Naudé (*MB* 3, 5, 7; 1727, 1737, 1743) handled the proof of the formula for the area of a quadrilateral inscribed in a circle (see I, p. 99) and the characterization of triangles by means of three suitable determining parts. Stewart (1746) and Simpson (1752) presented valuable collections of interesting theorems; Chapple (*Misc. Curiosa Math.* 2, 1746?) and Euler (*NCP* 11, 1765) set up the relationship $d^2 = r(r - 2\rho)$ for the triangle; Fuss (*NAP* 13, 1795–96), the corresponding relationship for the bicentric quadrilateral. Euler presented other propositions connected with particular points and lines of a triangle (*NCP* 11, 1765). The partition of figures was given by Tob. Mayer (mss. 1751, published in plagiarization by Wilke, 1757). Cramer's problem, in a given circle, to inscribe a triangle passing through three given points, arose in his development of material from Pappus (1742). Many solutions were found for this problem (Castillon-Lagrange: *NMB* 1776; Euler-Fuss-Lexell: *AP* 4, 1780), the prettiest of all, with an extension to the n-gon coming from A. Giordano, who was 16 years old at the time (1785; printed *Mem. Mat. Fis. Soc. Ital. Sc.* [= *MSI*] 4, 1788, Malfatti's solution there also). Constructions by means of a straight edge and a fixed circle were treated by Lambert (1774) and those by means of compasses alone by Mascheroni (1797, without knowledge of Mohr, 1672). Euler was the source of the interesting problem of the Königsberg bridges (*CP* 8, 1736) of the polyhedron theorem (*NCP* 4, 1752–53, without knowledge of Descartes as his predecessor) and of the question dealing with the number of ways in which a convex polygon can be partitioned into triangles by drawing diagonals (inadequately treated by Segner whose solution was corrected by Euler in *NCP* 7, 1758–59).

Trigonometry, heretofore merely an auxiliary science, now developed into an independent mathematical discipline. Fine compilations (Oppel 1746, Simpson 1748) were outstripped by Euler's excellent formal treatment of plane geometry (*Introd.* 1748) and to this he connected subsequent works: the construction of a spherical trigonometry, interpreting great circles as geodesic lines (*HMB* 9, 1753), the trigonometric determination of regular solids (*AP* 2, 1778) and a general spherical trigonometry (*AP* 3, 1779). Lambert (1765–70) placed great emphasis upon approximation formulas and transformations by means of hyperbolic functions which had already been introduced by V. Riccati (1757) in good logical order. The transition from spherical to plane trigonometry was carried out by Lambert (1765) and Legendre (*HMP* 1787). We are indebted to Lexell (*NCP* 19–20, 1774–75; *AP* 5–6, 1781–82), to Fuss (*NAP* 2, 10, 1784, 1792), to Schubert (*NAP* 3–4, 1785–86) and to L'Huilier (1799) for other individual contributions. An excellent summary of all that was known at the time was given by Cagnoli (1786). Finally, it lay within Lagrange's power to construct a unified system of trigonometric theory based exclusively upon the law of cosines (*JEP* 6, 1798–99). Mention should also be made of the tables by Lambert (1770), Schulze (1778), Hutton (1785), Vega (1783, 1794) and Callet (1795).

Algebra was a fixed part of the instruction in mathematics in the universities and in the secondary schools. The comprehensive introductory works mentioned earlier, without exception, also contained a course of study in algebra. There were, in addition, independent introductory works. The presentations by Saunderson (1740), Simpson (1745) and Maclaurin (1748) were held in high esteem not only in England, but also beyond its shores. Clairaut's introduction to Algebra (1746), which was constructed heuristically, exerted a revolutionary influence. Here, algebra was conceived of as a symbolic language, as it was later, by Condillac (1798) and by Condorcet (1799). Euler's guide to algebra, written in an exceptionally interesting manner (1770) was even more effective: editions with supplements by Lagrange on indeterminate analysis (from 1774) were especially valuable. In all these writings, the algorithmic point of view was given special prominence. Basic questions, somewhat like the matter of the difference between the sign before a term and the operational symbol, the systematic extension of the number concept, etc., were hardly touched upon and they had to be construed by students by way of their sense of feeling. There was no lack of critics who recognized these defects in the structure of the system, but there certainly was a lack of creative individuals who could find a remedy for this situation.

Advances in science were contained in individual essays which enjoyed rather small circulation and they were scattered in numerous current periodicals. In the first instance, the matter under discussion was the most appropriate partition of an expression into partial fractions in connection with problems of integration (De Moivre: 1730; Euler: 1748; Fuss: *AP* 1, 1777; Euler: *AP* 4, 1780). Then it was the proof of the Descartes rule of signs (Gua: *HMP* 1741; Lagrange *NMB* 1777), then, the elimination problem (Gua: 1740; Cramer: 1750; Euler: 1748; *HMB* 4, 20, 1748, 1764) and the solution of the system of linear equations arising here, through the employment of expressions of the determinant type (Cramer 1750; Bézout: *HMP* 1764 and 1779; Vandermonde: *HMP* 1772; Laplace: *HMP* 1772; Lagrange: *NMB* 1773). The first purely formal representation of the solutions of an equation in complex form led Euler to conjecture (*MB* 7, 1743) that every equation with real coefficients could be decomposed by real methods into linear and quadratic factors with real coefficients. To this, D'Alembert (*HMB* 2, 1746) joined a noteworthy attempt at proving the fundamental theorem of algebra. His effort was discussed by Gauss (1799) and completed by him at the critical point. Now for the first time it was established that elementary functions of a complex argument were complex (attempt at a proof by Euler: *HMB* 5, 1749; Foncenex: *MT* 1, 1759; Lagrange: *NMB* 1772). Kühn (*NCP* 3, 1750–51) was already trying to explain the multiplication rule for real and imaginary numbers geometrically. A serviceable representation of the complex numbers in a plane, upon which the more profound treatment of the subject depends, appeared for the first time in Wessel's work (1799). It passed unnoticed. Gauss, who was in possession of information equivalent to this from 1796, let the matter rest with allusions to it in 1799 and did not come forward with a more detailed presentation until 1831 (*Götting. gelehrte Anz.*).

In the course of the struggle over the general algorithmic solution of higher equations, equations of the third and fourth degrees were the first to be studied intensively (Euler: *CP* 6, 1732–33 and oftener, formation of resolvents; Bézout: *HMP* 1762; Waring: 1762; Lagrange: *NMB* 1770–71; Vandermonde: *HMP* 1771; Malfatti: *Atti. Ac. Siena* 1771; Mallet: 1782). Through the closely allied investigation of equations of the fifth degree, Malfatti (*Atti. Ac. Siena* 1771) found the resolvent of the sixth degree, and Bring achieved the transformation to the normal form $x^5 + px + q = 0$ (1786). Finally, the discovery of the types of equations which were solvable in radical form was made (Bézout: *HMP* 1762, 1765; G. Fr. Fagnano: 1770; Euler: 1776, printed *NAP* 6, 1788). The equations for the division of a circle were also investigated (Euler: *CP* 13, 1741–

43; Vandermonde: *HMP* 1771). These results were the starting point for Gauss's discovery of the constructibility of the regular polygon of 17 sides by means of compasses and straight edge (1796). The best compilations of the achievements to date came from Lagrange (*NMB* 1770–71) who, at that time, seriously doubted the possibility of finding a general algorithmic solution for higher equations. He also took the correct view of the structure of the solution for the case $n > 4$. By the introduction of hemisymmetrical functions, etc., he prepared the way for the procedure invented by Ruffini (1799: attempt at proving the impossibility of solving the equation of the fifth degree), by Abel (1824: *Crelles Journ.* 1, 1826: impossibility of solving the general equation of the fifth degree) and by Galois (from 1829: fundamental concepts of group theory).

Euler (*CP* 7, 1734–35) applied the so-called Newtonian formulas for the calculation of the power sums $s_k$ of the roots of an equation (first printing 1707, only for $k < n$) to the determination of the expression for $\Sigma 1 : k^s$ which had been the subject of unsuccessful research for more than 80 years. Proofs for the Newtonian formulas were produced by Bärmann (1745) and by Kästner (1757, printed in *Diss. math. phys. Soc. sc. Göttingen* 1771). The formulas for $k \geqq n$ were originated by Maclaurin (printed 1748); new derivations were given by Euler (1747; printed *Opusc.*, II, 1750). Independent formulas and the reduction of every other symmetrical function of the roots to power sums appeared in Waring (1762).

The trigonometric treatment of quadratic equations proved useful for numerical purposes (Halley in Kersey, 2nd ed. 1733; Simpson 1748; Dionis du Séjour 1786; Cagnoli 1786). Employing the same method, Cagnoli mastered equations of the third and fourth degree (last publication *MSI* 7, 1794). In addition to this, the method of recurring series was applied (Daniel Bernoulli *CP* 3, 1728; Euler: 1748; Lagrange: 1798). Other methods were also used: combinations involving the power sums of the roots of the equation (Lambert: *Acta Helvetica* 3, 1758; Lagrange: 1798), iteration procedures for trinomial equations (Lambert: *Acta Helv.* 3, 1758), and their development (Lagrange: *HMB* 24–25, 1770–71; Lambert: *NMB* 1770; Euler: *AP* 3, 1779 and often), the use of equations having roots equal to squares of the differences of the roots of the original equation (Waring: *PT* 53, 1763) and the continued fraction method (Lagrange: *HMB* 23–24, 1767–68, 1794–95; 1798). The graphical representation of formulas of higher degree through suitably drawn straight lines may be traced back to Segner (*NCP* 7, 1758–59). This method served the purpose of ascertaining a favorable initial value for the approximation of the roots of an equation. Rowing extended this procedure instrumentally (*PT* 60, 1770).

Many of the subjects touched upon here had already passed beyond the elementary domain and they were in the nature of a transition to questions which could be approached only by the specialists of the day who possessed the highest qualifications.

## 2. *Leading Personalities on the Continent*

The leading mathematicians of the age of enlightenment were decisively influenced by Jean Bernoulli and his school, which saw in Euler its greatly admired standard bearer. L. Euler (1707–1783) was the oldest son of the country clergyman, Paul Euler, who during his own student days had become interested in mathematics and for that reason had attended Jacques Bernoulli's lectures. The wide awake boy received his first instruction from his father. Then, he attended the Basle gymnasium (in bad condition at the time) and along with this, he became a private pupil of Burckhardt. In 1720, Euler entered the Basle university and, as a fellow student of Daniel and Nicolas (II) Bernoulli, under Jean Bernoulli, he made rapid progress. He earned the degree of *Bacc.* in 1721 and the degree of *Mag. Art.* in 1723, made friends with Jean Bernoulli's son, Jean (II) Bernoulli, broke off the theological studies which had been selected for him in favor of mathematics exclusively and by 1725, he had already handled algebraic reciprocal trajectories with great elegance (printed: *AE* 1727).

Because of his youth, Euler was not admitted as a candidate for the position of professor of physics which had become vacant at Basle (1727), and he accepted the invitation sent him by Nicolas (II) Bernoulli and Daniel Bernoulli to come to the Petersburg Academy (founded 1725) where both men were teaching. There, he came in contact with the widely travelled East Prussian, Goldbach, who had emigrated to Petersburg with Hermann in 1725, and who, in 1742, rose to be Russian minister of state. Euler was appointed professor of physics at the Academy in 1730. In 1733, after Daniel Bernoulli had returned to Basle, Euler became his successor as the Academy professor of mathematics.

Euler's first scientific works in Petersburg arose entirely from the areas of thought over which the Bernoulli school ranged. Most valuable of all was the *Mechanica* (1736), leading the trend in this field. His excellent arithmetic texts (printed 1738–40) also came from the same period. They were used in the gymnasium connected with the Academy—a model school where Euler acted as examiner and Krafft was a teacher. Krafft, in Petersburg until 1744, wrote an interesting elementary geometry for this school (1740) and later (1753) he presented a valuable historical survey of elementary methods in higher geometry. The chief subject of discussion at the Petersburg Academy was summation of series. Goldbach transformed series by term by term comparison (*CP* 2, 1727); Daniel Bernoulli employed recurrent series in the approximation method of solving equations (*CP* 3, 1728). Euler represented terms of a series by definite integrals; he used $\int_0^1 x^p (1 - x)^n \, dx$ (*B*-function) and $\int_0^1 x^p e^{-x} \, dx$ (Γ-function) in imaginative interpolations, passing from positive integral values of the argument $p$ to fractional values, and he made use of fractional indices as in differentiation (*CP* 5, 1730–31). Through bold generalizations, Euler arrived at the summation value of $\pi^2 : 6$. This work was independent of Stirling who in 1730, by means of $\sum\limits_{k=n}^{\infty} 1 : k^2 = \sum\limits_{k=1}^{\infty} \dfrac{1}{k^2} : \binom{n + k - 1}{k}$ and $n = 13$, obtained the value of $\sum\limits_{k=1}^{\infty} 1 : k^2$ first to 8 and then to 17 correct decimal places. Jean Bernoulli guessed Euler's procedure from the mere statement of his results (*Op.* IV, 1742); Jean Bernoulli, himself, employed the additional relationship $\pi^2 : 8 = \int_0^1 arc \sin x \, dx : \sqrt{1 - x^2}$ (by letter 1737, printed: *Journ. litt. d'Allemagne* 2, 1743). Nicolas (I) Bernoulli (*CP* 10, 1738) gave a numerical proof by means of the Leibniz series through the use of another special case of the Euler transformation (see p. 90), which he simplified later (1750 notes). One of Euler's early studies was on the subject of geodesic lines on a surface expressed in space coordinates (*CP* 3, 1728). Euler's attention was directed to Fermat's results in the theory of numbers by Goldbach, author of the well known conjecture in number theory (letter to Euler of June 7, 1742). Through a study of the Fermat works, Euler came to the conclusion that not all numbers of the form $2^p + 1$ (with $p = 2^q$) were prime numbers (examples in counterevidence: $q = 5$) (*CP* 6, 1732–33). Euler studied the equation $x^2 - py^2 = 1$ (CP 6), proved the lesser Fermat theorem $a^{p-1} \equiv 1$ (mod. $p$) (CP 8, 1738) and the impossibility of $a^4 + b^4 = c^2$ in positive integers (CP 10, 1738). During the same period, he discovered the formula named after him for the sum of semiconvergent series (CP 6; proof CP 8; independent proof by Maclaurin also, 1742). Through an investigation of harmonic series (CP 7, 1734–35 and 8, 1736), he determined the Euler constant $\frac{1}{1} + \frac{1}{2} + \frac{1}{3} + ... + \frac{1}{n} - ln(n+1)$. He developed several approximation methods for the calculation of $\pi$ through series (CP 9, 11, 12, 1737, 1739, 1740), among which the method employing the addition theorem for the arc tangent function was included.

During the same period, Euler began his studies on continued fractions, from which the irrationality of $e$ and $e^2$ proceeded directly (CP 9, 11, 1737, 1739). He worked on infinite products (CP 11) and on the isoperimetric problem (CP 6, 1732–33 and 8, 1736). The germ of almost every great result of Euler's later period was laid in the Petersburg years. Euler deliberately turned away from the synthetic method which was still zealously cultivated in England, replacing it with the analytical methods. Notwithstanding that these results made a celebrity of Euler, his stay in Petersburg was a joyless one during which he was vexed by petty restraints imposed by the board of directors of the Academy. For that reason, in 1741 he returned to Berlin where, in 1745, he became the director of the mathematics division of the Berlin Academy.

Clairaut, the highly gifted son of a Parisian mathematician, was another member of the Bernoulli circle of friends. A. Cl. Clairaut (1713–1765) studied the work of Guisnée (1705) when he was nine years old. At the age of ten, he studied L'Hospital's work on conic sections (1707) and then his *Analyse* (1696). In

1726, he presented a study on curves of the fourth order before the *Ac. Sc.* (printed: *MB* 4, 1734), and he worked on space curves and surfaces. In 1731, a study on the determination of the centroids of space forms and a treatise on space curves which had been completed in 1729, appeared in the *HMP*. The latter work procured a membership in the *Ac. Sc.* for the 18-year-old author by a circumvention of the regulations providing for an age limit. At almost the same time, Jac. Hermann wrote on equations of surfaces (*CP* 6, 1732–33).

The *HMP* of 1733 presented Clairaut's theorems on surface curves having extreme properties and geodesic lines on surfaces of revolution; the *HMP* of 1734 presented his investigations on singular solutions of systems of curves which were given by their differential equations. Clairaut's close connection with Maupertuis began at this time. In 1729, Maupertuis and König, together, had studied under Jean Bernoulli. In the fall of 1734, Maupertuis travelled to Basle with Clairaut for the purpose of discussing preparations for the expedition to Lapland (1736–37). This expedition resulted in a decision concerning the form of the terrestrial ellipsoid in favor of the Newtonian concept and against that of Descartes. In the *HMP* of 1739, Clairaut treated the integrating factor and he proved Nicolas (I) Bernouilli's "axiom" (1721), that $\frac{\partial^2 z}{\partial x \, \partial y} = \frac{\partial^2 z}{\partial y \, \partial x}$. In 1740, he worked on homogeneous equations. The scientific evaluation of the Scandinavian expedition led him (*HMP* 1739) to the investigation of geodesic lines on almost spherical surfaces of revolution; it led, in 1743, to a treatise on the form of the earth (level surfaces of a power function, first start in potential theory), in 1752–54, to the construction of a useful theory of the moon. His introduction to geometry (1741) and algebra (1746), previously mentioned, enjoyed an extraordinarily good reception.

Maupertuis, who had been corresponding with Frederick the Great ever since 1731, was installed as president of the Academy upon the latter's accession to the throne. The Academy was greatly in need of reform, but the planned reorganization did not go into effect until after the cessation of both Silesian Wars (1745). Euler (1744) had just placed the calculus of variations on a new foundation through an appropriate utilization of integration by parts (Euler equation). Like Maupertuis, who had offered the presentation and critical analysis of the theory of monads as a subject for a prize competition at the academy, Euler, also, was an opponent of the Leibniz philosophy. However, he understood this philosophy only superficially and at second hand. Consequently, his attempts to refute it on the basis of primitive-mechanistic reasoning, which even then, was behind the times, were made in a dilettantish manner (1746, despite general opposition, repeated many times 1768). Euler undertook the defense (*Opusc.* I, 1746) of the principle of least action enunciated by Maupertuis, on the basis of which the latter proposed to prove the existence of God (*HMP* 2, 1746; 1750). He embraced the president's cause against the critical objections put forward by König (*NAE* of Mar. 1751). Apart from other matters, König's objections had arisen when he noted the mention of the Maupertuis principle in a letter by Leibniz dated Oct. 17, 1707. Only a copy of the letter had been made available and this was to an unknown recipient. Euler declared the letter to be a forgery (*HMB* 6, 1750). König stood to his own defense in an *Appel au Public* (1752). Following this, Voltaire's satirical tale of Dr. Akakia (1752) made not only Maupertuis, but Euler, as well, the laughing stock of all Europe. Euler's efforts (1753) to save the situation were in vain. Maupertuis's health failed and in 1756 he retired from the presidency which remained vacant thereafter. Even though Euler took over the administration of the Academy, he never achieved a personal relationship with the king. The king was basically prejudiced against mathematics and would admit its value only on the grounds of its usefulness. After the close of the Seven Years' War, the king offered the presidency to D'Alembert. The latter declined out of consideration for Euler, who had put through the appointment of several Swiss scholars to the Academy. Lambert was one of these (1764). Highly talented though he was, Lambert was completely unfamiliar with academic conventions. In 1766 unendurably offended by Frederick's harsh behavior toward him, Euler accepted a tempting call to Petersburg tendered to him by Katherine II.

Euler's great textbooks were written during his stay in Berlin. They were exciting to read, skillfully presented and they were widely published in many editions. The popularization of infinitesimal methods began with these books. His two volume *Introductio* (1745, printed 1748) led the way. The first volume began with algebraic analysis, including the theory of number series, power series, recurrent series and continued fractions. Now, logarithmation was introduced as a second inversion of raising to a power. Exponential and logarithmic series were obtained from the binomial series by passing to a limit; trigonometric functions were no longer defined as line segments but rather as ratios between line segments. The Euler identity, inverse trigonometric series, real composition of conjugate complex functions and the representation of integral functions as infinite products, followed in quick succession. Euler placed great value upon rapidly converging expansions. The second volume contained plane analytical geometry including the discussion of curves and a concise survey of algebraic and transcendental curves. In addition, solid analytical geometry was given, including the theory of surfaces of the second order and the treatment of space curves as sections of a projecting cylinder.

In its fundamentals, the differential calculus (1748, printed 1755) was not satisfactory (differentials were defined as null quantities), but algorithmically, it was superlative. It contained, for example, the Euler transformation of series, the recursive representation of the Bernoulli numbers, the reversal of series, the treatment of indeterminate expressions and the theory of interpolation. In the three volume work on the integral calculus (1763, printed 1768–70) integration was defined as the inverse of differentiation; the definition of integration as a process of summation was completely lacking. Formal techniques of integration of rational, irrational and elementary transcendental functions now followed and these were followed in turn by the treatment of differential equations of the first and higher orders through the method of separation of variables, through the application of an integrating factor, and through series expansion. Allied to the treatment of singular solutions, some material was given on total differential equations, on the calculus of variations, on the differential equation of the elliptical integral and on transcendental functions such as the integral logarithm.

Mathematical research produced, above all, proofs of Fermat's theorems in number theory—in ways, to be sure, which were not those of Fermat. The main achievements were the proofs for the possibility of presenting numbers in the form $4n + 1 = a^2 + b^2$ (*NCP* 1, 3, 5, 1747–48, 1750–51, 1754–55), the determination of pairs of integers which would satisfy $x^2 - py^2 = 1$ (1759; printed: *NCP* 11, 1765) and the proof of the impossibility for $a^3 + b^3 = c^3$ (1759; printed: *NCP* 8, 1760–61). In addition to this, trigonometric interpolation (*MB* 7, 1743 and often), and the solution of differential equations by series and by definite integrals with parameters were given. These were followed by profound works on elliptic integrals (from 1754, leading to the addition theorem for the three kinds). Euler also treated the vibration of a string (*HMB* 4, 9, 1748, 1753; in conflict with D'Alembert, *HMB* 3, 6, 1747, 1750, and with Daniel Bernoulli *HMB* 9), the construction of formulas for the rotation of a rigid body (*HMB* 14, 1758), the series expansion of the differential equation of cylindrical functions (1761; printed: *NCP* 10, 1764), the solution of partial differential equations (from 1762) and the totality of the radii of curvature of a right section of a surface (1763; *HMB* 16, 1760). On the other hand, Euler failed to come up to expectations in the formal handling of divergent series which he represented by closed expressions obtained by means of inadmissible expansion. The justifiable objections raised to this by Daniel Bernoulli and by Nicolas (I) Bernoulli (by letter 1741–43) were not taken into account because the results obtained by inadmissible reasoning could, nevertheless, be proved correct for the most part by procedures of various types, and consequently Euler pronounced them as free of objections, on an intuitive basis at least.

Shortly after his emigration to Petersburg, Euler, who had lost his right eye as early as 1735, suffered complete blindness from a cataract. He found selfless and devoted helpers in his oldest son, J. Al. Euler and particularly (from 1773) in the young man from Basle, N. Fuss (pupil of Daniel Bernoulli and Euler's future grandson-in-law). These two made it possible for the patriarch who had already become legendary, to edit great numbers of additional scientific works of which approximately 250 were on pure mathematics. The first work dictated after he had become blind, was the guide to algebra, mentioned previously (1767, printings since 1768). The indeterminate exercises included there, which were to be treated for the most part merely as individual cases, were supplemented by numerous more difficult problems of the same type. To this he added the reciprocity law for quadratic residues (1772; *Opusc. Anal.* I, 1783) and the construction of the *numeri idonei* (from 1778). In 1777, Euler discovered the equivalence of the Brouncker continued fraction expansion for 4 : $\pi$ and the Leibniz series (*NAP* 2, 1784). Almost simultaneously, he advanced to Fourier series (*NAP* 11, 1793) and on this basis, he developed a new determination of $\Sigma 1 : k^2$ (*Opusc. Anal.* I, 1783). From 1768 on, Euler treated the double integral·(*NCP* 14, 1769) and in 1771 he took up Lagrange's method of variation (*NCP* 16). In connection with Lambert's investigations in cartography (*Beytrage* III, 1772), Euler worked on area-preserving and angle-preserving mapping of the surface of a sphere on a plane (1775; *AP* 1, 1777).

Jean Bernoulli had already sought for rectifiable curves on the surface of a sphere (*Op.* III, 1742). In 1770, Euler handled the same subject (*NCP* 15, 1770). In 1771, he presented rectifiable curves on the cone of revolution (*AP* 5, 1781) and on surfaces of revolution of the second order (*NAP* 3, 1785). Then he gave algebraic curves having equal arcs (*NAP* 4, 1786) and such curves as were related to the parabola or the ellipse through the equality of arcs (*NAP* 5, 1787). In connection with partial differential equations, Euler added studies of canals and cornice areas (*NCP* 14, 1769), on developable surfaces (*NCP* 16, 1771) and on the fundamental properties of space curves in differential geometry (1775; *AP* 6, 1782). In 1777, Euler presented an elegant method of determining the ellipse through four and correspondingly through three given points, which would enclose the maximum area (*NAP* 9, 1791), and in 1782, shortly before his sudden death, he was working on orthogonal families of curves (*MP* 7, 1815–16).

Euler's honors lie principally in the algorithmic domain, not in fundamentals. Through the abundance and the diversity of the methods which were at his command, despite bold and debatable methods of reasoning, he attained remarkable results, which, for the most part, were proved rigorously for the first time in the 19th century. It is true that he was disappointed in his hope that higher equations would prove to be solvable in radicals, but Euler did at least discover interesting families of curves (1776). His adherence to Jean Bernoulli's concept of a function was surprising. This restricted the domain of functions to those which were characterized by the exclusive use of operational symbols. Here Euler fell just short of the mark (Fourier series). Taking the differentials as equivalent to zero was unfortunate; so also was his definition of integration as the inverse of differentiation. Without the summation definition, the way to an understanding of line integrals and the like was blocked. On the other hand, Euler prepared for the triumph of the analytical methods of thinking over synthetic methods which were tied all too tightly to subject matter, and for those who followed him, he opened an approach to a field of activity of unforeseen extent.

Lambert, likewise, produced important results in the field of theoretical as well as practical matters. J. H. Lambert (1728–1777) born in Mulhouse, came of a Huguenot emigrant family. His father was a poor tailor. Young Lambert had been outstanding even when attending the town school, but he could not continue to study there because of the lack of funds. He worked as his father's assistant and at the same time he educated himself through books which came his way by chance. Later on, in his capacity as a librarian, then as a private secretary and after that as a resident tutor (1748–58), he invented many practical devices (e.g., perspectograph: mss. 1752) and he worked (like Bilfinger, follower of Leibniz, who taught at the Petersburg gymnasium 1725–31) on an algebraic logic (printed 1764 and 1771). His distinguished work on perspective (1759) was also useful on the level of the practical man (constructions in the unfavorable circumstances). It constituted an important advance over Taylor's work (revised editions of 1749, 1755, 1757) which up to this time, had been the leading work in the field. As early as 1766, Lambert's work was taken into consideration by Zanotti. The appendices to his original work added by Lambert in 1774 contained constructions by straight edge and a fixed circle, and also historical notes which were influenced by Montucla's history of mathematics (1758). In 1761, his *Cosmology* appeared. This work was independent of the Kantian theory (1755), but nevertheless it was related to it in its fundamental ideas (developed by Lambert as early as 1748).

In his correspondence (1765–66), Lambert opposing Euler, characterized the infinitely small as a pure fiction. Two years later, he demonstrated by continued fraction expansion that $e^x$ and $tg\ x$ were irrational for all rational values of $x$. By taking $x = \pi : 4$, it followed that $\pi$ was also irrational. In the *Beyträge* I (1765) there were noteworthy ideas on methods in trigonometry. The basic formulas for the right spherical triangle were derived from a single formula by the use of the *pentagramma mirificum*, in continued development of methods due to Vieta and to Napier (see I, p. 93) and in anticipation of the group theory point of view.

Lambert's theory of parallel lines was written in 1766 (printed: *Mag. M.* 1786). Here he established the existence of an absolute measure of length and the proportionality of the area of a surface to its excess or deficiency, on a surface of constant curvature $\neq 0$. Lambert conjectured that the geometry of the sphere had a corresponding counterpart on an "imaginary sphere." In 1769, Lambert placed trigonometric functions and hyperbolic functions on a par as aids. He employed them in the representation of imaginary quantities (*HMB* 1768; printed: 1770) and he computed a small table of hyperbolic functions (1670).

In his contributions to the theory of equations of 1768 (*HMB*

1763; printed: 1770) Lambert expressed serious doubts concerning the possibility of an algorithmic solution of higher equations. He developed the basic trigonometric formulas in the theory of quadrilaterals (*Beyträge* II), worked on questions in probability (*NMB* 1771), treated (*Beyträge* III) stereographic projection and the problem of interpolation, and presented rapidly converging expansions of elliptic integrals. As early as 1762, Lambert endeavored to obtain the greatest possible rigor in the treatment of infinitesimal questions (printed: *Mag. M.* 1788) but he encountered an absence of comprehension on the part of most of his contemporaries. In his view, practical applicability was to be greatly emphasized in theoretical researches.

No scientist of his time was a match for Euler in abundance of ideas and in the power of organization, but yet, in fundamentals and in rigorous methods of thinking, he was substantially surpassed by the highly gifted Lagrange. J. L. Lagrange (1736–1813) was the son of an official of French descent in the Turin commissariat, who had suffered severe losses in hazardous speculations. It was the father's desire to see his first born devote himself to the lucrative profession of the law. However, while the young man was studying at the secondary school under Ravelli, he developed a taste for geometry. Applying himself completely to mathematics, he studied the great works which had just been made available through publication in print, by such authors as Jean Bernoulli (1742), by Jacques Bernoulli (1744), by Newton (1744). He also studied Leibniz's correspondence with Jean Bernoulli (1745) and manuscripts by Leibniz and Euler. He engaged in a correspondence with G. C. Fagnano (1754) and with Euler (1755). At the early age of 19, he was appointed professor at the Turin artillery school.

Lagrange was one of the founders of the Turin Academy

(1757) in whose *Miscellanea Taurinensia* (= *MT*) his first works appeared. In volume I, (1759) extreme values of functions of several variables, linear differential equations and the problem of the transverse vibration of a string were treated. Volume II (1760–61) contained work fundamental to the calculus of variations (application of the δ-method, sent to Euler by letter as early as 1755), other examples of practical applications, the differential equation of minimal surfaces and of developable surfaces, and a study on the principle of least action. In addition to this, there were series expansions for the integration of ordinary linear differential equations and the simpler partial differential equations, and a fundamental consideration in the field of infinitesimal processes, namely, that such processes are admissible only if the errors committed in passing to a limit are neutralized. In volume III, (1762–65) there were approximate methods for the solution of differential equations and observations on simultaneous systems and on total differential equations. In volume IV, (1766–69) methods in the calculus of variations were improved. In the investigation of linear differential equations of higher order, adjoint differential equations were set up, and in the formal solution, the variation of the constants was employed. Lagrange avoided the use of the integrating factor in the treatment of the Euler differential equation of the elliptic integral. He also presented the first rigorous proof for the continued fraction solution of $x^2 - py^2 = 1$. Volume V (1770–73) contained interesting studies on errors in observation.

In consequence of the excessive mental exertions involved in the editing of the works of his youth, Lagrange suffered an illness which became apparent in 1761. It brought on an attack of neurasthenia, and it could be compensated for only by a strictly regulated way of life. As conditions in Turin were unsatisfactory, Lagrange—a corresponding member of the Berlin Academy ever since 1757—stepped into the vacancy left by Euler (1766). His great works in algebra were written in Berlin. The numerical solution (including the continued fraction method) which appeared in the *HMB* 23–25, 1767–69 (collected 1798) was supplemented by the studies on the algebraic solution through an extension of the group theory point of view (*NMB* 1770–77; also applied in the 1797 revision of the Clairaut algebra by Lacroix). Added to this, there was the treatment of the equation $x^2 - py^2 = q$ by continued fractions and the characterization of the divisor of $ax^2 + bxy + cy^2$. The Lagrange reversal formula for general functions originated in the extension of the Lambert approximation procedure for the solution of trinomial equations (*HMB* 24, 1768; proved by Rothe, through combinatorial processes, *Arch. M.* 1, 1794). Beyond this, he formulated a symbolical differential and integral calculus, and the method for finding unknown periods. Lagrange also wrote a memorandum on the nature of the infinitesimal processes from which it appears that he had begun to entertain some doubts concerning his earlier explanation (*NMB* 1772). The next volume of the *NMB* contained the transformation of multiple integrals, the presentation of distinguishing magnitudes of a tetrahedron based on its six edges, and also the introduction of curvilinear coordinates. This was followed by the theory of singular solutions of ordinary differential equations with prospects of corresponding considerations in partial differential equations (*NMB* 1774) and the solution of the equation $tg\ y = \dfrac{1+m}{1-m}\ tg\ x$ in the form $y = x + m\sin 2x + \dfrac{1}{2}\ m^2 \sin 4x + \dfrac{1}{3}\ m^3 \sin 6x\ + \ \ldots$ (1774, printed *NMB* 1776). In the *NMB* of 1775 the subjects of discussion were problems in probability, characteristic partial differential equations of the second order for problems in physics and recurrent series. In this work, it was established that an inadvertent error had slipped by Malfatti (*MSI* 3, 1786–87) and this was corrected by Lagrange (*NMB* 1792–93). The *NMB* of 1779 contained other studies on partial differential equations, and on the mapping of a spherical surface on a plane. In 1784, at Lagrange's suggestion, the Berlin Academy took the nature of infinity as the subject of a prize competition. The prize was won by L'Huilier (1786) with a well balanced presentation, placing the question of passing to a limit at the center of the discussion. Carnot also expressed himself along similar lines (not submitted, first published 1797). Lagrange's treatment of the arithmetico-geometrical means, likewise came in 1784 (*Mém. T.* (2) 2, 1784–85).

After the death of Frederick the Great (1786) Lagrange went to Paris as a pensioned member of the *Ac. Sc.* (1787). He became a naturalized citizen of France and his epoch making *Mécanique* was published (1788). His general equations of motion, derived from D'Alembert's principle (1743) and also the series expansion for the solution of $\frac{\partial^2 \varphi}{\partial x^2} + \frac{\partial^2 \varphi}{\partial y^2} + \frac{\partial^2 \varphi}{\partial z^2} = 0$ were given here. Another treatise was concerned with the problem of solids of revolution (1788). Lagrange was a member of the commission for the reformation of the system of weights and measures in 1790–91. He published his interpolation formula in the *NMB* 1792–93. However, it was not until 1795 that he was once again intensively occupied with mathematics. He taught for a short time at the newly organized *École Normale* (lectures printed 1812) and then he entered the newly established *Institut de France* as its leading member. From 1797 on, Lagrange taught at the *École Polytechnique*. In connection with the Taylor expansion, whose remainder he investigated in rather great detail, Lagrange tried to treat the infinitesimal processes by purely

algebraic methods (1797 and 1799). In 1801, he presented his
conclusions on singular solutions of differential equations and
in the foreword to the second edition of his *Mécanique* (1808)
he emphasized the heuristic value of all considerations in the
infinitesimal domain.

Gifted with surpassing intuitive power, Monge became the
creator of descriptive geometry. He placed analytical geometry
of space, and differential geometry on a new foundation, and he
amalgamated the analytical and the geometrical methods in the
treatment of partial differential equations. G. Monge (1746–
1818) was the highly talented son of a hardworking knife grinder
and shopkeeper of Piedmontese extraction. He was lovingly educated by the Oratorians of his native town of Beaune; at
their recommendation, he became a teacher of physics at the
Lyons College of the Oratorians (1762–64), but he did not
desire to enter their Order. Because of his station in life, he
was trained at the engineer's school in Mézières not as an officer
candidate but rather as a foreman. Having attracted attention to
himself by the construction of a new type of plan for fortifica-
tions, he was given first the position of coach under Bossut who
was professor of mathematics at Mézières and then, in 1768, the
position of professor as Bossut's successor. He was a successful
teacher there, lecturing on descriptive geometry. The methods
for this work were evolved by Monge in an ingenious development of beginnings made by Derand (1634), Jousse (1642) and
the presentation in Frézier's work (1737–39 and 1760) which
had already advanced beyond these. However, for military reasons, he was not permitted to publish these methods before 1794.
At that time he published his masterful lectures given at the
*École Normale* (*Géom. Descr.*) and at the *École Polytechnique*
which he had founded and directed (Feuilles).

As early as 1768, Monge worked on problems in variations

along the same lines as Euler and Lagrange (several works printed in the *Mém. T.* 5, 1770–73). From 1771 on, he handled basic problems of solid analytical geometry and advanced as far as the metrical line coordinates named after Plücker (*PT* 155, 1865). At the same time, he developed the fundamentals of differential geometry of space curves and of the developable surfaces and evolutes connected with them. The study on differential geometry (1771; printed: *MSP* 9, 1780) by his oldest pupil (1769–71) and scientific friend, Tinseau, was written in the spirit of Monge. In this work, there appeared, among other things, the osculating plane of a space curve, the plane tangent to a surface and the tangent cone from a point to a surface. In the winter of 1771–72, Monge submitted to the *Ac. Sc.* two treatises (unprinted at that time) on special linear differential equations of the first order which he solved by arbitrary functions (also "discontinuous", i.e. represented in various domains by expressions of various types) and which he related to the determination of the surface of a family through a given curve. In 1772, other examples followed (printed *Mém. T.* 5, 1770–73; *MSP* 7, 1776), and then, in succession, there were: in 1773, a (lost) comprehensive manuscript on determinants, in 1774, the treatment of functional equations (printed: *MSP* 9), in 1775, a study on developable surfaces, their characteristic strips between neighboring generators and on the boundaries of shadows and penumbras in the illumination of an object by a luminous body (printed: *MSP* 9), in 1776, results obtained in lines of curvatures and focal surfaces (printed: *HMP* 9). These were supplemented by the treatise by Meusnier (Monge's pupil 1774–75) on the normal curvature of a surface curve (1776; printed *MSP* 10, 1785), on the osculating paraboloid and on minimal surfaces (in addition to the first examples).

Monge, who had been a corresponding member of the *Ac. Sc.*

since 1772, was admitted as a member of the *Ac. Sc.* in 1780. This did not prevent him from carrying out his duties as professor at Mézières. At the same time he also gave private instruction in analytical and infinitesimal methods to the almost destitute Lacroix (1781–82) in Paris. The latter took over many of Monge's basic concepts in the widely used parts of his *Cours* (from 1795). In 1783, Monge, as Bézout's successor, became examiner of the midshipmen. For a rather long time, he worked almost exclusively in the fields of physics and chemistry. It was 1785–86 before he again submitted new mathematical treatises to the *Ac. Sc.* (printed: *HMP* 1784 and *Mém. T.* (2) 1, 1784–85). They contained the reduction of linear partial differential equations of the first order to simultaneous systems, they achieved the reduction of non-linear equations to linear equations and gave examples of tangential transformations. In addition to this, there was a study suggested by Lagrange's essays, on singular solutions of differential equations and, in particular, on the family of lines represented by the equation $y - px = f(p)$.

During the French Revolution, Monge fought passionately against the *ancien régime*, serving for a short period as minister of naval affairs (1793–94). He, together with his former pupil, Carnot (1771–73) organized the *Levée en masse* in 1794. On an official mission for the *Directoire* in 1796, he became acquainted with Napoleon and in 1798–99, he participated in the Egyptian expedition. During the First Consulate, he was called to the senate and he was overwhelmed with honors. Nevertheless, upon the overthrow of Napoleon in 1816, he was excluded from the *Ac. Sc.* He died in 1818 after a long illness.

Most of the studies on differential geometry and the theory of surfaces were collected and unified in the *Feuilles* of 1794–95 and in the numerous enlarged editions of it. Monge's pupil, Hachette was the principal participant in these efforts. Here, the method of characteristics in the solution of partial differential equations and the formulation of lines of curvature for the triaxial ellipsoid had their first publication. On the other hand, the use of line coordinates disappeared in the third edition (1807), probably for pedagogical reasons. We can do no more than skim lightly over Monge's elegant lesser contributions in the *Journal de l'École Polytechnique* ( = *JEP*) and in the *Corresp. de l'École Polytechnique*. Among them, there was a study in the *JEP* 15, 1809 in which the formula for the triangle and the tetrahedron in Cartesian coordinates came up and where the algebraic sign before a term was correctly interpreted.

The theory of geometric transformations also began with Monge. His work dealt with the polar of a point in space with respect to a pair of straight lines, given (about 1790) and with the relationship between reciprocal polars with respect to a paraboloid of revolution (*Géom. Descr.*). The theorem that the radical axes of three circles meet in a point arose in the lectures given at the *Éc. Polyt.* (reported by Poncelet, 1822). How much more Monge contributed to this field by way of inspiration through personal contact, we do not know. Through his profound works, he created the foundations for the development of geometrical methods in the modern sense. These had their beginnings in the studies by his pupil, Brianchon (*JEP* 13, 1806: Brianchon hexagon) and by Poncelet (1822).

Whereas Monge's principal scientific achievement fell within the period of the *ancien régime* and merely the publication of it and the consequences of its publication ran over into the revolutionary years, the creative life work of Laplace and of Legendre belong only in small part to the closing days of the rococo period.

P. S. Laplace (1749–1827) came of a well-to-do Norman family of the aristocracy. He studied in Caen from 1765, and from 1772, he taught at the Paris military academy where

Napoleon was numbered among his students (1784–85). During his brief term of office (1795) as minister of the interior under the Consulate, Laplace failed to live up to expectations, but despite this, he was overwhelmed with honors bestowed upon him by Napoleon. Nevertheless, in 1813, he went over to the side of Louis XVIII and he was raised to the peerage of France by the latter in 1817.

In the field of mathematics, Laplace devoted himself, above all, to questions of probability (*HMP* 1774 and often; *MSP* 6, 1775 and often; collected 1812 and often). He made a detailed critical analysis of the views of Daniel Bernoulli (*CP* 1730–31 and often: mathematical and moral expectation, Petersburg problem; *HMP* 1760 and often: application of infinitesimal methods to the theory of probability), of the doubts expressed by D'Alembert concerning the interpretation hitherto prevailing (*Enc.* 1754 and often) and above all, of the work of Bayes (*PT* 53–54; 1763–64). Bayes used the expression $\frac{2}{\sqrt{\pi}} \int_0^u e^{-u^2} \, du$ to determine the probability of the occurrence of an event, so that the probability in favor of the occurrence would lie between the limits $\frac{p}{p+q}\left\{ 1 \pm u \cdot \sqrt{\frac{2q}{p(p+q)}} \right\}$ for the case of very large numbers of favorable occurrences ($p$ times) and likewise of failures to occur ($q$ times). This led Laplace in his considerations of probability, to the use of recurrent series for the determination of $\int_0^\infty e^{-u^2} du$ (*HMP* 1779) and the valuable results in the summation of difference equations (*HMP* 1779). Laplace also made important contributions to the treatment of partial differential equations (cascade method: *HMP* 1773 and *MSP* 6, 1774).

He handled the Lagrange equation $\dfrac{\partial^2 \varphi}{\partial x^2} + \dfrac{\partial^2 \varphi}{\partial y^2} + \dfrac{\partial^2 \varphi}{\partial z^2} = 0$

through the application of spherical functions (HMP 1782–83) and on this occasion, he also went into the determination of the numerical values produced by the multiplication of a great many fractions expressed in large numbers. The elementary introductory lectures which he gave at the *École Normale* were noteworthy works (1794–95, printed: *JEP* 2, 1812). Laplace's main achievement was the *Traité de Mécanique Céleste* (1799–1825), which grew out of decades of study of individual questions in astronomy (from 1773). It contained, for example, the expansion of functions in spherical functions, the treatment of confocal surfaces of the second order and many other matters of mathematical interest.

A. M. Legendre (1752–1833) was educated at the *Collège Mazarin*. In consequence of D'Alembert's intercession on his behalf, he taught beside Laplace at the Paris military academy from 1775 to 1780. In 1783, he was accepted as a member in the *Ac. Sc.*, in 1787 he was entrusted with geodetic problems, and in 1791 he became a member of the commission for the reformation of weights and measures. Legendre taught at the *École Normale* from 1795, and he became Laplace's successor as examiner at the *Éc. Polyt.* in 1799.

Legendre investigated the polynomial named after him in 1784 and he applied it to the representation of functions (*HMP* 1784, 1789; *MSP* 10, 1785). He discovered the reciprocity law for quadratic residues (*HMP* 1785, independently of Euler) and later on, he expressed it in the Legendre symbolism (1797–98). He introduced the second variation into the calculus of variations and he stated the criterion for great extremes (*HMP* 1786). He treated (*HMP* 1787) partial differential equations through the application of the Legendre tangential transformations and geodesic lines to surfaces of revolution of the second order. In connection with his profound, if not altogether successful, studies on the parallel axiom, he taught the calculation of a spherical triangle which was very nearly a plane triangle, by a method of approximation, keeping the lengths of the sides fixed and diminishing each angle by one third of the spherical excess. In the *HMP* of 1786, Legendre made the rectification of the arc of a hyperbola depend upon the rectification of a pair of elliptical arcs (method similar to Landen's, *PT* 61, 1771, of whose work he was not yet aware). By means of his original procedures, he confirmed the validity of the Landen transformation of elliptic integrals (*PT* 65, 1775; reprinted 1780) and the determination of isoperimetric ellipses (semiaxes of one: *a, b;* of the other: $b \pm e$). That this result, together with the calculation of the perimeter of the ellipse by iteration was an accomplishment already to the credit of Jean Bernoulli (see p. 94, 95) had passed into oblivion. Legendre's great work on elliptic transcendents was written in 1792 (printed 1794; known for the first time in foreign countries through the reprint of 1811). It contained the systematic treatment of the integral $\int \Re [x, \sqrt{f_4(x)}] dx$, its distinct subdivision into three kinds and their mutually interchangeable relationships, the addition, multiplication and division formulas.

In his widely read *Éléments de Géométrie* (1794 and very often) Legendre tried to achieve a reorientation toward Euclid, turning away from the genetic method and yet at the same time applying algebraic methods of reasoning. He was not uninfluenced by Simson, but he took too little account of the desiderata regarding method expressed by contemporaries like D'Alembert who possessed deeper insight into fundamentals. Individual questions were treated in supplements, as in the case of Lexell's theorem (*AP* 1781, 1st ed.) stating that the vertices of all spherical triangles of equal area, having the same base, lie on a small circle parallel to the base, the theory of inscribed quadrilaterals which was very simply formulated through the application of algebraic expedients, and also the proof of the irrationality of $\pi$ and $\pi^2$ by continued fraction expansion, which closed a gap in a proof by Lambert.

Legendre's number theory (from 1797–98) contained the (not completely correct) conjecture that the total number of primes up to $n$ was approximately $n : (ln\ n - 1.08366)$. It also contained a copious, but not very methodical survey of the pertinent results of the older school. The presentation of the adjustment of observations by the method of least squares (1805, given without proof) came after its appearance in studies by Gauss (from 1795, printed 1809). Although the numerous contributions to the integral calculus (1811–17) and to the theory of elliptical integrals (1825–1832) brought with them many new points of view, they must still be regarded primarily as excellent collections embodying the theories developed in the 18th century.

The Revolution did, indeed, bring the *ancien régime* to an end, but it had no such effect upon the tendencies of the enlightenment. These grew stronger during the Revolution. In the Napoleonic period, they made even greater progress. They found expression in Laplace's concept of the mechanistic determination of world events (1814). From now on, new and comprehensive theories unfolded in diversified formal procedures, and insofar as this obtained, the French Revolution represented a significant turning point even in the realm of mathematics.

## 3. Addenda

To fill out the history of the era, several other fields of interest in mathematics during the period of enlightenment should be treated. They lay at the center of the total discussion somewhat less forcefully than those already mentioned.

To begin with, reference should be made to the continued development of the Newtonian "organic" generation of curves through the rotation of a rigid angle. Maclaurin (*PT* 30, 1719; then 1720), using the methods of pure geometry, found $\frac{1}{2}(n-1)(n-2)$ to be the number of double points possible for a curve of the $n$th order. He determined these curves by $\frac{1}{2}n(n+3)$ points and he knew that a $C_m$ and a $C_n$ intersect in $mn$ points. Maclaurin also rediscovered Pascal's theorem which, meanwhile, had been completely forgotten, and he gave a systematic treatment of pedal curves. Braikenridge (1726; printed: 1733) who knew nothing of Maclaurin's writings, replaced the angle with a straight line, e.g. he rotated the straight lines $a, b, c$ about the fixed points $A, B, C$ respectively, so that $b \times c$ and $c \times a$ described specified curves, and he investigated the locus of $a \times b$. The outstanding general theorems originated by Maclaurin (from 1721) on the general properties of diameters and points of intersection, first appeared as supplements to his posthumously printed algebra (1748).

Maupertuis made noteworthy observations (*HMP* 1729) on the manner in which points of inflection and cusps follow upon one another on a finite branch of a curve. He showed also, that when these points move toward coincidence, they lead to higher singularities. Bragelonge (*HMP* 1730–32) added several discoveries on asymptotes having the form of straight lines and curves, and he undertook a classification of curves of the fourth order according to their configuration (known only from a short account). Gua (1740) transformed Newton's analytical parallelogram into a triangle and was thus enabled to indicate infinitely distant singularities. Nevertheless, he restricted himself almost exclusively to the use of Cartesian coordinate geometry and he mentioned the type of curves analysis employed in differential geometry only incidentally. In her very painstakingly written textbook (1748), where, with scrupulous care Agnesi presented the algebraic and infinitesimal methods known at that time, numerous examples in the differential geometry of curves were given in addition to many examples of algebraic curves. Cramer's *Introductio* (1750) was the best resumé of the subject of algebraic curves. It contained a detailed presentation and analysis of all the results achieved up to 1740, corrected Gua's inaccurate observations on singularities (cusps of the second kind), carried out the series expansion of a branch of a curve, brought in the ratio of curvatures and directed attention in a particularly impressive manner to the Cramer paradox regarding the number of points required for the determination of curves of the $n$th order. In 1756, Dionis du Séjour and Goudin issued a fresh compilation of results obtained up to that time. Through the addition of an observation to the effect that a curve of the $n$th order can have at most $n(n-1)$ tangents having the same slope, they prepared for the concept of classes (Poncelet, *Ann. math.* 1818).

Waring (1762) constructed unique extreme value theorems for algebraic curves, discovered the invariance of order under projection and coined the concept of a surface of the $n$th order. In addition, he observed that no algebraic curve having an oval and no double points was quadrable by means of elementary functions. In the less original compilation by Riccati and Saladini (1765–67), special attention should be given to the well thought out examples of the application of the infinitesimal calculus to plane curves. For special curves, reference must once more be made to Loria (from 1902).

Several textbooks on infinitesimal mathematics must also be

listed. They could not, indeed, measure up to those of Euler or Lagrange, but as solid technical works they were popular. Besides the previously mentioned collections by Agnesi (1748), by Riccati and Saladini (1765–67), there were books written by Leseur and Jacquier (1768), Cousin (1777), Vega (1782) and Wydra (1783).

Buffon contributed an interesting monograph on the calculus of probability (report: *HMP* 1733; in full 1777). His question regarding the probability that a circular disc will fall completely within one compartment of a square grid, and related questions (needle problem) touched upon the difficult field of geometrical probability for the first time.

In England, where the hold of the synthetic method was of longest duration, noteworthy research was undertaken in the reconstruction of the writings of Apollonius. Horsely (1770), Lawson (1771), Wales (1772), Simson (posthumously 1776) and Burrow (1779) worked in this field. We are likewise indebted to the English mathematicians for the publication of the distinguished edition of Archimedes by Torelli (1792, posthumously).

The most important mathematicians in England during the second half of the 18th century were Landen and that remarkable personality, Ed. Waring (1734–1798). In 1762, Waring independently arrived at formulas for the power sums of the roots of an equation, and on the basis of these, he represented all the symmetrical functions of the roots. He placed the roots between limits by the use of appropriate functions of the differences of the roots, and approximated their values by determining the equation which was satisfied by the squares of the roots of the original equation and continuing accordingly (anticipation of the ideas of the Gräffe method: 1837–39). Waring also surmised the insolvability of the general equation of the fifth degree by means of radicals. In 1770, he presented a new statement of

Goldbach's empirical theorem, and also the so-called Wilson theorem, $(p - 1)! \equiv - 1 \pmod{p}$, which had been known to Leibniz and for which Euler had given an elegant proof (*Opusc. anal.* I, 1783). Beyond this, Waring asserted that every number is the sum of at most $f(p)$ $p$th powers (first proof by Hilbert, *Math. ann.* 67, 1909). Waring also treated convergence of series and the conditions for convergence (1776 and *PT* 74, 1784), and in 1779 (*PT* 60) he enunciated the so-called Lagrange interpolation formula, although his applications of it were not as many-sided as those of the latter.

Finally, there is the unique Hindenburg school of combinatorial analysis to be considered. In 1778–79, Hindenburg proceeded from the investigations of the multinomial theorem and its extension to infinite series. The high point of the studies, carried out (in impractical symbolism) by this group, was Eschenbach's reversal of series (1789) which Rothe put into a form free of objections (1793). In a misjudgment of the facts, Hindenburg believed that "combinatorial" methods were on a par with algebraic and infinitesimal methods. The school soon collapsed because its adherents failed to produce additional results at a moment when the great French mathematicians were at the peak of their productivity, developing their ideas in easily comprehended expositions of dynamic force and attraction.

It is, above all, to the credit of the devoted efforts of Lagrange, Monge, Laplace and Legendre, their coworkers and pupils in the fields of research and teaching, that the individual personalities interested in mathematics became ever increasing in number, first in France and then over the whole of Europe. Regrouping in the social order consummated in consequence of the French Revolution, growing industrialization and technical advances led, in connection with new thoughts on the ways and means of education, its significance and procedures, to a complete reorganization of the educational system. Teachers in the universities attained full freedom of instruction. Possibilities of effective scientific activity, unknown up to that time, were opened to them. The study of special subjects was encouraged and perfected. In the field of mathematics, there was a complete change in fundamental concepts. From this time on, there was a striving for the greatest possible rigor in the use of algorithmic methods and toward clear, distinct and simple methods of notation in demonstrations. Purely intuitive observations were rolled back in favor of well founded logical methods of reasoning and the evolution of general and comprehensive points of view. By slow degrees, these ideals and aspirations prevailed.

# Index

155